*Cancer Diagnostics with
DNA Microarrays*

Cancer Diagnostics with DNA Microarrays

Steen Knudsen

Medical Prognosis Institute, Hørsholm, Denmark

WILEY-LISS

A John Wiley & Sons, Inc., Publication

Copyright © 2006 by John Wiley & Sons, Inc. All rights reserved.

Published by John Wiley & Sons, Inc., Hoboken, New Jersey.
Published simultaneously in Canada.

For general information on our other products and services or for technical support, please contact our
Customer Care Department within the United States at (800) 762-2974, outside the United States at (317)
572-3993 or fax (317) 572-4002.

Wiley also publishes its books in a variety of electronic formats. Some content that appears in print may
not be available in electronic formats. For more information about Wiley products, visit our web site at
www.wiley.com.

Library of Congress Cataloging-in-Publication Data is available.

ISBN-13: 978-0-471-78407-4
ISBN-10: 0-471-78407-9

Printed in the United States of America.

10 9 8 7 6 5 4 3 2 1

To Tarja

Contents

Preface

A new technology is about to enter cancer diagnostics. DNA microarrays are currently showing great promise in all the medical research projects to which they are being applied. This book presents the current status of the area as well as reviews and summaries of the results from specific cancer types. For types where several comparable studies have been published, a meta-analysis of the results is presented.

This book is intended for a wide audience from the practicing physician to the statistician. Both will find the chapters where I review areas of their expertise superficial, but I hope they will find other introductory chapters useful in understanding the many aspects of microarray technology applied to cancer diagnostics.

I first describe the current state of the technology as well as emerging technologies. Then I describe the statistical analysis that is necessary to interpret the data. Next I cover some of the major human cancer types where microarrays have been applied with success, including studies that I have been a part of. I conclude that for several cancer types the results are so good and consistent that DNA microarrays are ready to be deployed in clinical practice. The clinical application in question is helping to select patients for adjuvant chemotherapy after surgery by determining the prognosis more accurately than what is possible today.

Chapters 1–6 on technology and statistical analysis and Chapters 9 and 10 on chip design and software are updated and expanded versions of chapters in my previous Wiley book, *Guide to Analysis of DNA Microarray Data, Second Edition* (2004). The remaining 13 chapters are new.

STEEN KNUDSEN

Birkerød, Denmark
October 2005

Acknowledgments

My previous books, on which some of the chapters in this book are based, were written while I worked for the Technical University of Denmark. I am grateful to the University leadership, the funding agencies, my group members, and collaborators for their significant role in that endeavor. The review of individual cancer types, to a large extent based on information from the American Cancer Society, was also written during my employment at the Technical University of Denmark.

The remaining chapters were written while I was employed by the Medical Prognosis Institute. The original research on meta-analysis of cancer classification as well as subnetwork mapping of individual cancer types has been patented by Medical Prognosis Institute, and I am grateful for being allowed to present the results here.

S. K.

1

Introduction to DNA Microarray Technology

1.1 HYBRIDIZATION

The fundamental basis of DNA microarrays is the process of *hybridization*. Two DNA strands hybridize if they are complementary to each other. Complementarity reflects the Watson–Crick rule that adenine (A) binds to thymine (T) and cytosine (C) binds to guanine (G) (Figure 1.1). One or both strands of the DNA hybrid can be replaced by RNA and hybridization will still occur as long as there is complementarity.

Hybridization has for decades been used in molecular biology as the basis for such techniques as Southern blotting and Northern blotting. In Southern blotting, a small string of DNA, an *oligonucleotide*, is used to hybridize to complementary fragments of DNA that have been separated according to size in a gel electrophoresis. If the oligonucleotide is radioactively labeled, the hybridization can be visualized on a photographic film that is sensitive to radiation. In Northern blotting, a radiolabeled oligonucleotide is used to hybridize to messenger RNA that has been run through a gel. If the oligo is specific to a single messenger RNA, then it will bind to the location (*band*) of that messenger in the gel. The amount of radiation captured on a photographic film depends to some extent on the amount of radiolabeled probe present in the band, which again depends on the amount of messenger. So this method allows semiquantitative detection of individual messengers.

DNA arrays are a massively parallel version of Northern and Southern blotting. Instead of distributing the oligonucleotide probes over a gel containing samples of RNA or DNA, the oligonucleotide probes are attached to a surface. Different probes can be attached within micrometers of each other, so it is possible to place many of them on a small surface of one square centimeter, forming a DNA array. The sample is labeled fluorescently and added to the array. After washing away excess unhybridized material, the hybridized material is excited by a laser and is detected by

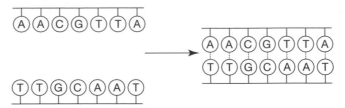

Figure 1.1 *Hybridization of two DNA molecules. Dotted line: hydrogen bonds.*

a light scanner that scans the surface of the chip. Because you know the location of each oligonucleotide probe, you can quantify the amount of sample hybridized to it from the image generated by the scan.

There is some contention in the literature on the use of the word "probe" in relation to microarrays. Throughout this book the word "probe" will be used to refer to what is attached to the microarray surface, and the word "target" will be used to refer to what is hybridized to the probes.

Where before it was possible to run a couple of Northern blots or a couple of Southern blots in a day, it is now possible with DNA arrays to run hybridizations for tens of thousands of probes. This has in some sense revolutionized molecular biology and medicine. Instead of studying one gene and one messenger at a time, experimentalists are now studying many genes and many messengers at the same time. In fact, DNA arrays are often used to study *all* known messengers of an organism. This has opened the possibility of an entirely new, systemic view of how cells react in response to certain stimuli. It is also an entirely new way to study human disease by viewing how it affects the expression of all genes inside the cell. I doubt there is any disease that does not in some way affect the expression of genes in some cells. That is the basis for this book. By applying DNA microarrays to human tissue or human cells, we can learn about disease and characterize disease at a much more detailed level than what was previously possible. Cancer has turned out to be the disease that has attracted the most focus. One reason is that it is possible to obtain a tissue sample from the tumor during surgery. This tissue sample is then used to measure gene expression with a DNA microarray.

DNA arrays can also be used to hybridize to DNA. In that case they can distinguish between different alleles (mutations) in DNA, some of which affect cancer. DNA chips have been developed for detecting mutations in the human *TP53* tumor suppressor gene (Ahrendt et al., 1999; Wikman et al., 2000; Spicker et al., 2002). Likewise, a DNA chip for detecting mutations in the human *CYP450* genes, important for metabolizing common drugs, has been developed. These genotyping or resequencing chips will not be covered further in this book, which focuses exclusively on gene expression in cancer.

More recently, arrays have been used for comparative genomic hybridization (CGH) to reveal deletion or duplication of chromosomal regions in cancer (Pollack et al., 2002; Douglas et al., 2004; Blaveri et al., 2005a; Jones et al., 2005). While the results have been very promising, they are limited in number and will not be covered in this book. Similarly, arrays have recently been used to detect small noncoding RNAs (miRNA) and a correlation to cancer has been found (Lu et al., 2005). This will not be covered further in this book.

1.2   THE TECHNOLOGY BEHIND DNA MICROARRAYS

When DNA microarrays are used for measuring the concentration of messenger RNA in living cells, a *probe* of one DNA strand that matches a particular messenger RNA in the cell is used. The concentration of a particular messenger is a result of *expression* of its corresponding gene, so this application is often referred to as *expression analysis*. When different probes matching all messenger RNAs in a cell are used, a snapshot of the total messenger RNA pool of a living cell or tissue can be obtained. This is often referred to as an *expression profile* because it reflects the expression of every single measured gene at that particular moment. Expression profile is also sometimes used to describe the expression of a single gene over a number of conditions.

Expression analysis can also be performed by a method called *serial analysis of gene expression* (SAGE). Instead of using microarrays, SAGE relies on traditional DNA sequencing to identify and enumerate the messenger RNAs in a cell (see Section 1.2.7).

Another traditional application of DNA microarrays is to detect mutation in specific genes. The massively parallel nature of DNA microarrays allows the simultaneous screening of many, if not all, possible mutations within a single gene. This is referred to as *genotyping*.

The treatment of array data does not depend so much on the technology used to gather the data as it depends on the application in question.

For expression analysis the field has been dominated in the past by two major technologies. The Affymetrix, Inc. GeneChip system uses prefabricated oligonucleotide chips (Figures 1.2 and 1.3). Custom-made chips use a robot to spot cDNA, oligonucleotides, or PCR products on a glass slide or membrane (Figure 1.4).

More recently, several new technologies have entered the market. In the following, several of the major technology platforms for gene expression analysis are described.

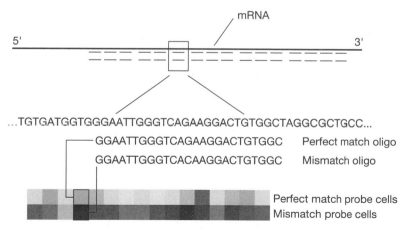

Figure 1.2 *The Affymetrix GeneChip technology. The presence of messenger RNA is detected by a series of probe pairs that differ in only one nucleotide. Hybridization of fluorescent messenger RNA to these probe pairs on the chip is detected by laser scanning of the chip surface. (Figure by Christoffer Bro.)*

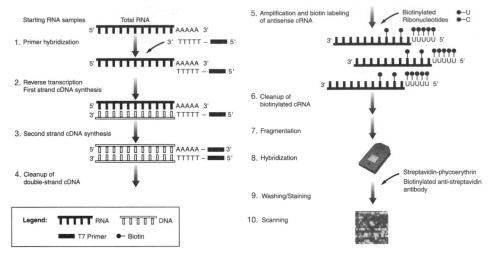

Figure 1.3   *Preparation of sample for GeneChip arrays. Messenger RNA is extracted from the cell and converted to cDNA. It then undergoes an amplification and labeling step before fragmentation and hybridization to 25-mer oligos on the surface of the chip. After washing of unhybridized material, the chip is scanned in a confocal laser scanner and the image is analyzed by computer. (Image courtesy of Affymetric)*

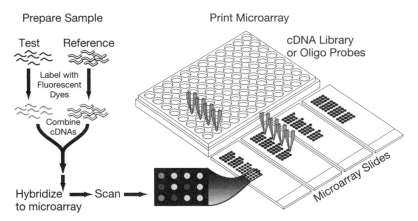

Figure 1.4   *The spotted array technology. A robot is used to transfer probes in solution from a microtiter plate to a glass slide where they are dried. Extracted mRNA from cells is converted to cDNA and labeled fluorescently. Reference sample is labeled red and test sample is labeled green (see color insert). After mixing, they are hybridized to the probes on the glass slide. After washing away unhybridized material, the chip is scanned with a confocal laser and the image is analyzed by computer.*

1.2.1   Affymetrix GeneChip Technology

Affymetrix uses equipment similar to that which is used for making silicon chips for computers, and thus allows mass production of very large chips at reasonable cost. Where computer chips are made by creating masks that control a photolithographic process for removal or deposition of silicon material on the chip surface, Affymetrix uses masks to control synthesis of oligonucleotides on the surface of a chip. The standard phosphoramidite method for synthesis of oligonucleotides has been modified to allow light control of the individual steps. The masks control the synthesis of several hundred thousand squares, each containing many copies of an oligo. So the result is several hundred thousand different oligos, each of them present in millions of copies.

That large number of oligos, up to 25 nucleotides long, has turned out to be very useful as an experimental tool to replace all experimental detection procedures that in the past relied on using oligonuclotides: Southern, Northern, and dot blotting as well as sequence specific probing and mutation detection.

For expression analysis, up to 40 oligos are used for the detection of each gene. Affymetrix has chosen a region of each gene that (presumably) has the least similarity to other genes. From this region 11 to 20 oligos are chosen as perfect match (PM) oligos (i.e., perfectly complementary to the mRNA of that gene). In addition, they have generated 11 to 20 mismatch (MM) oligos, which are identical to the PM oligos except for the central position 13, where one nucleotide has been changed to its complementary nucleotide. Affymetrix claims that the MM oligos will be able to detect nonspecific and background hybridization, which is important for quantifying weakly expressed mRNAs. However, for weakly expressed mRNAs where the signal-to-noise ratio is smallest, subtracting mismatch from perfect match adds considerably to the noise in the data (Schadt et al., 2000). That is because subtracting one noisy signal from another noisy signal yields a third signal with even more noise.

The hybridization of each oligo to its target depends on its sequence. All 11 to 20 PM oligos for each gene have a different sequence, so the hybridization will not be uniform. That is of limited consequence as long as we wish to detect only *changes* in mRNA concentration between experiments. How such a change is calculated from the intensities of the 22 to 40 probes for each gene will be covered in Section 3.3.

To detect hybridization of a target mRNA by a probe on the chip, we need to label the target mRNA with a fluorochrome. As shown in Figure 1.3, the steps from cell to chip usually are as follows:

- Extract total RNA from cell (usually using TRIzol from Invitrogen or RNeasy from QIAGEN).
- Separate mRNA from other RNA using poly-T column (optional).
- Convert mRNA to cDNA using reverse transcriptase and a poly-T primer.
- Amplify resulting cDNA using T7 RNA polymerase in the presence of biotin-UTP and biotin-CTP, so each cDNA will yield 50 to 100 copies of biotin-labeled cRNA.
- Incubate cRNA at 94 degrees Celsius in fragmentation buffer to produce cRNA fragments of length 35 to 200 nucleotides.
- Hybridize to chip and wash away nonhybridized material.
- Stain hybridized biotin-labeled cRNA with streptavidin-phycoerythrin and wash.
- Scan chip in confocal laser scanner (optional).
- Amplify the signal on the chip with goat IgG and biotinylated antibody.
- Scan chip in scanner.

Usually, 5 to 10 µg of total RNA are required for the procedure. But new improvements to the cDNA synthesis protocols reduce the required amount to 100 ng. If two

TABLE 1.1 Performance of the Affymetrix GeneChip Technology.

	Routine Use[a]	Current Limit[a]
Starting material	5 μg total RNA	2 ng total RNA
Detection specificity	1×10^5	1×10^6
Difference detection	Twofold changes	10% changes
Discrimination of related genes	70–80% identity	93% identity
Dynamic range (linear detection)	Three orders of magnitude	Four orders of magnitude
Probe pairs per gene	11	4
Number of genes per array	47,700	47,700

[a] Numbers refer to chips in routine use and the current limit of the technology (Lipshutz et al., 1999; Baugh et al., 2001).

cycles of cDNA synthesis and cRNA synthesis are performed, the detection limit can be reduced to 2 ng of total RNA (Baugh et al., 2001). MessageAmp kits from Ambion allow up to 1000 times amplification in a single round of T7 polymerase amplification. The current performance of the Affymetrix GeneChip technology is summarized in Table 1.1.

1.2.2 Spotted Arrays

In another major technology, spotted arrays, a robot spotter is used to move small quantities of probe in solution from a microtiter plate to the surface of a glass plate. The probe can consist of cDNA, PCR product, or oligonucleotides. Each probe is complementary to a unique gene. Probes can be fixed to the surface in a number of ways. The classical way is by nonspecific binding to polylysine-coated slides. The steps involved in making the slides can be summarized as follows (Figure 1.4):

- Coat glass slides with polylysine.
- Prepare probes in microtiter plates.
- Use robot to spot probes on glass slides.
- Block remaining exposed amines of polylysine with succinic anhydride.
- Denature DNA (if double-stranded) by heat.

The steps involved in preparation of the sample and hybridizing to the array can be summarized as follows (Figure 1.4):

- Extract total RNA from cells.
- Optional: isolate mRNA by polyA tail.
- Convert to cDNA in the presence of amino-allyl-dUTP (AA-dUTP).
- Label with Cy3 or Cy5 fluorescent dye linking to AA-dUTP.
- Hybridize labeled mRNA to glass slides.
- Wash away unhybridized material.
- Scan slide and analyze image (see example image in Figure 1.5).

The advantage compared to Affymetrix GeneChips is that you can design any probe for spotting on the array. The disadvantage is that spotting will not be nearly

Figure 1.5 *Spotted array containing more than 9000 features. Probes against each predicted open reading frame in Bacillus subtilis are spotted twice on the slide. Image shows color overlay after hybridization of sample and control and scanning (see color insert). (Picture by Hanne Jarmer.)*

as uniform as the in situ synthesized Affymetrix chips and that the cost of oligos, for chips containing thousands of probes, becomes high. From a data analysis point of view, the main difference is that in the cDNA array usually the sample and the control are hybridized to the same chip using different fluorochromes, whereas the Affymetrix chip can handle only one fluorochrome so two chips are required to compare a sample and a control. Table 1.2 shows the current performance of the spotted array technology.

Applied Biosystems (www.appliedbiosystems.com) has recently developed a chemiluminescent microarray analyzer with increased sensitivity. The recommended starting material is 1 µg of total RNA, but as little as 100 ng of total RNA is possible.

TABLE 1.2   Performance of the Spotted Array Technology (Schema, 2000)

	Routine Use
Starting material	10–20 µg total RNA
Dynamic range (linear detection)	Three orders of magnitude
Number of probes per gene	1
Number of genes or ESTs per array	~40,000

1.2.3 Digital Micromirror Arrays

In 1999, Singh-Gasson and co-workers published a paper in *Nature Biotechnology* showing the feasibility of using digital micromirror arrays to control light-directed synthesis of oligonucleotide arrays. Two commercial companies were formed based on this technology. NimbleGen (www.nimblegen.com) synthesizes DNA arrays using digital micromirrors and sells the manufactured arrays to the customer. Febit (www.febit.de) makes an instrument, the Geniom One (see Table 1.3), which allows the customer to control digital micromirror synthesis in his or her own lab (Baum et al., 2003). The design of the microarray is uploaded by computer and the synthesis takes about 12 hours in a DNA processor (Figures 1.6 and 1.7). Then the fluorescently labeled sample is added, and after hybridization and washing the fluorescence is read by a CCD camera and the resulting image is returned to the computer. All steps are integrated into a single instrument.

1.2.4 Inkjet Arrays

Agilent has adapted the inkjet printing technology of Hewlett Packard to the manufacturing of DNA microarrays on glass slides. There are two fundamentally different approaches. Presynthesized oligos or cDNAs can be printed directly on a glass surface. These are called deposition arrays. Another approach uses solid-phase phosphoramidite chemistry to build the oligos on the array surface one nucleotide at a time.

TABLE 1.3 Performance of the Febit Geniom One Technology (Febit, 2005)

Starting material	5 μg total RNA
Detection limit	0.5 pM spiked transcript control
Number of probes per gene	1–10
Probe length	User selectable (10–60 mers)
Feature size	34 μm by 34 μm
Probes per array	Maximum 15,000
Arrays per DNA processor	Up to eight
Sample throughput	About 80 samples per week

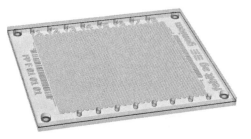

Figure 1.6 *The Febit DNA processor with microchannel structure. (Left) A two-dimensional view of the microchannels. (Right) A three-dimensional view of the microchannels including inlet and outlet of each channel. (Copyright Febit AG. Used with permission.)*

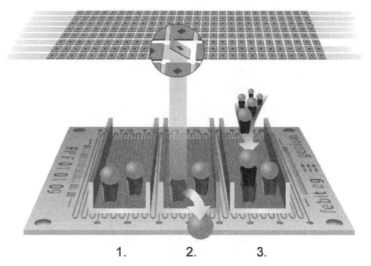

Figure 1.7 *Graphical illustration of the in situ synthesis of probes inside the Febit DNA processor. Shown are three enlargements of a microchannel, each illustrating one step in the synthesis (see color insert): 1 — the situation before synthesis; 2 — selected positions are deprotected by controlling light illumination via a micromirror; 3 — substrate is added to the microchannel and covalently attached to the deprotected positions. (Copyright Febit AG. Used with permission.)*

1.2.5 Bead Arrays

Illumina (www.illumina.com) has marketed a bead-based array system (see Table 1.4). Instead of controlling the location of each spot on a slide, they let small glass beads with covalently attached oligo probes self-assemble into etched substrates. A decoding step is then performed to read the location of each bead in the array before it is used to hybridize to fluorescently labeled sample.

1.2.6 Electronic Microarrays

Several companies have taken the merger between DNA and computer chip one step further and placed the DNA probes on electronically addressable silicon chips. Nanogen (www.nanogen.com) has made such a NanoChip Electronic Microarray, where binding is detected electronically. Combimatrix (www.combimatrix.com) has also made an electronically addressable semiconductor chip that allows the control of DNA synthesis at each site. A CustomArray Synthesizer is available for chip synthesis in the customer's lab. Binding is detected via conventional fluorescence.

TABLE 1.4 Performance of Bead Arrays (Illumina, 2003)

Starting material	50–200 ng total RNA
Detection limit	0.15 pM
Number of probes per gene	4–10
Probe length	50 mers
Feature spacing	6 µm
Probes per array	1500
Array matrix	96 samples

1.2.7 Serial Analysis of Gene Expression (SAGE)

A technology that is both widespread and attractive because it can be run on a standard DNA sequencing apparatus is serial analysis of gene expression (SAGE) (Velculescu et al., 1995; Yamamoto et al., 2001). In SAGE, cDNA fragments called tags are concatenated by ligation and are sequenced. The number of times a tag occurs, and is sequenced, is related to the abundance of its corresponding messenger. Thus, if enough concatenated tags are sequenced, one can get a quantitative measure of the mRNA pool. Bioinformatics first enters the picture when one wishes to find the gene corresponding to a particular tag, which may be only 9 to 14 bp long. Each tag is searched against

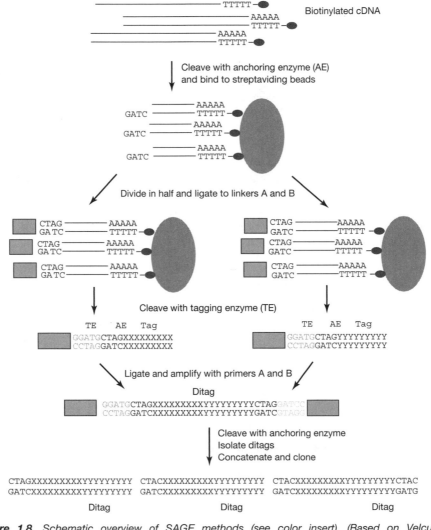

Figure 1.8 Schematic overview of SAGE methods (see color insert). (Based on Velculescu et al., 1995.)

a database (Lash et al., 2000; Margulies and Innis, 2000; van Kampen et al., 2000) to find one (or more) genes that match.

The steps involved in the SAGE methods can be summarized as follows (see also Figure 1.8):

- Extract RNA and convert to cDNA using biotinylated poly-T primer.
- Cleave with a frequently cutting (4-bp recognition site) restriction enzyme (anchoring enzyme).
- Isolate 3′-most restriction fragment with biotin-binding streptavidin-coated beads.
- Ligate to linker that contains a type IIS restriction site for and primer sequence.
- Cleave with tagging enzyme that cuts up to 20 bp away from recognition site.
- Ligate and amplify with primers complementary to linker.
- Cleave with anchoring enzyme and isolate ditags.
- Concatenate and clone.
- Sequence clones.

The analysis of SAGE data is similar to the analysis of array data described throughout this book except that the statistical analysis of significance is different (Audic and Claverie, 1997; Lash et al., 2000; Man et al., 2000).

1.3 PARALLEL SEQUENCING ON MICROBEAD ARRAYS

A conceptual merger of the SAGE technology and the microbead array technology is found in the massively parallel signature sequencing (MPSS) on microbead arrays, marketed by Solexa (www.solexa.com). cDNA is cloned into a vector together with a unique sequence tag that allows it to be attached to a microbead surface where an antitag is covalently attached (Figure 1.9). Instead of quantifying the amount of attached cDNA to each bead, the number of beads with the same cDNA attached is determined. This is done by sequencing 16–20 base pairs of the cDNA on each

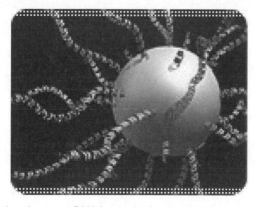

Figure 1.9 *Megaclone bead arrays. cDNA is attached to bead surface via tag–antitag hybridization. (From Solexa. Used with permission.)*

bead. This is done by a clever procedure of repeated ligation and restriction cycles intervened by fluorescent decoding steps. The result is the number of occurrences of each 16–20 bp signature sequence, which can be used to find the identity of each cDNA in a database just as is done with SAGE.

1.3.1   Emerging Technologies

Nanomechanical cantilevers (McKendry et al., 2002) can detect the hybridization of DNA without any fluorescent labeling. The cantilevers are made of silicon, coated with gold, and oligonucleotide probes are attached. When target–probe hybridization occurs, the cantilever bends slightly and this can be detected by deflection of a laser beam. The amount of deflection is a function of the concentration of the target, so the measurement is quantitative. An alternative to laser beam detection is piezo-resistive readout from each cantilever. As the number of parallel cantilevers increases, this technology shows promise for sensitive, fast, and economic quantification of mRNA expression. The company Concentris (www.concentris.com) currently offers a commercial eight-cantilever array (see Figure 1.10).

But nanotechnology ultimately will allow us to detect single molecules. When that happens, all the intermediate steps we currently use to amplify and label transcripts will become obsolete. At that time, nanotechnology-based detection of mRNA transcripts may replace the methods currently in use. But it may be five or ten years from today. Another possibility is that recognition of proteins improves so much that it can compete with DNA microarrays.

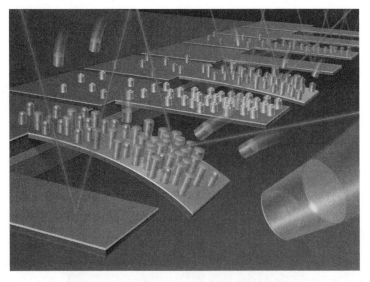

Figure 1.10   *Artist's illustration of array of eight nanomechanical cantilevers. Binding of targets leads to bending that is detected by deflection of a laser beam (see color insert). (From Concentris. Used with permission.)*

1.4 SUMMARY

Spotted arrays are made by deposition of a probe on a solid support. Affymetrix chips are made by light mask technology. The latter is easier to control and therefore the variation between chips is smaller in the latter technology. Spotted arrays offer more flexibility, however. Data analysis does not differ much between the two types of arrays. Digital micromirror technology combines the flexibility of the spotted arrays with the speed of the prefabricated Affymetrix chips.

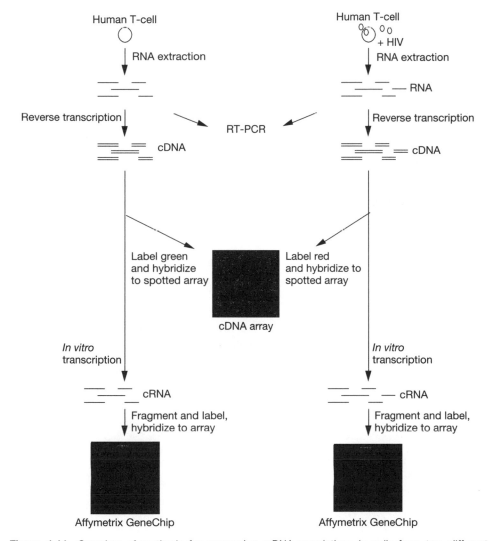

Figure 1.11 *Overview of methods for comparing mRNA populations in cells from two different conditions (see color insert).*

TABLE 1.5 Overview of Commercially Available Microarray Technologies

Technology	Factory Synthesis	Customer Synthesis
Affymetrix	Mask-directed photolithography	
Agilent	Inkjet	
NimbleGen	Micromirror photolithography	
Febit		Micromirror photolithography
Combimatrix		Electronically addressed synthesis
Spotted arrays	Robot spotting	Robot spotting
MPSS	Cloning and sequencing	
Solexa Beads	Oligos attached to beads	

Serial analysis of gene expression (SAGE) is yet another method for analyzing the abundance of mRNA by sequencing concatenated fragments of their corresponding cDNA. The number of times a cDNA fragment occurs in the concatenated sequence is proportional to the abundance of its corresponding messenger.

Figure 1.11 shows a schematic overview of how one starts with cells in two different conditions (e.g., with and without HIV virus) and ends up with mRNA from each condition hybridized to a DNA array. (See Table 1.5.)

FURTHER READING

Hood, L. (2002). Interview by The O'Reilly Network (http://www.oreillynet.com/).

Overview and Details of Affymetrix Technology

Affymetrix (1999). *GeneChip Analysis Suite*. User Guide, version 3.3.

Affymetrix (2000). *GeneChip Expression Analysis*. Technical Manual.

Lockhart, D. J., Dong, H., Byrne, M. C., Follettie, M. T., Gallo, M. V., Chee, M. S., Mittmann, M., Wang C., Kobayashi, M., Horton, H., and Brown, E. L. (1996). Expression monitoring by hybridization to high-density oligonucleotide arrays. *Nature Biotechnol.* 14:1675–1680.

Wodicka, L., Dong, H., Mittmann, M., Ho, M. H., and Lockhart, D. J. (1997). Genome-wide expression monitoring in *Saccharomyces cerevisiae. Nature Biotechnol.* 15:1359–1367.

Overview and Details of Spotted Arrays

Bowtell, D., and Sambrook, J. (editors) (2002). *DNA Microarrays: A Molecular Cloning Manual.* New York: Cold Spring Harbor Laboratory Press.

Pat Brown lab web site (http://brownlab.stanford.edu).

Schena, M. (1999). *DNA Microarrays: A Practical Approach* (Practical Approach Series, 205). Oxford, UK: Oxford University Press.

Microarrays web site (http://www.microarrays.org/). Includes protocols largely derived from the Cold Spring Harbor Laboratory Microarray Course manual.

Digital Micromirror Arrays

Albert, T. J., Norton, J., Ott, M., Richmond, T., Nuwaysir, K., Nuwaysir, E. F., Stengele, K. P., and Green, R. D. (2003). Light-directed $5' \rightarrow 3'$ synthesis of complex oligonucleotide microarrays. *Nucleic Acids Res.* 31(7):e35.

Beier, M., Baum, M., Rebscher, H., Mauritz, R., Wixmerten, A., Stahler, C. F., Muller, M., and Stahler, P. F. (2002). Exploring nature's plasticity with a flexible probing tool, and finding new ways for its electronic distribution. *Biochem. Soc. Trans.* 30:78–82.

Beier, M., and Hoheisel, J. D. (2002). Analysis of DNA-microarrays produced by inverse in situ oligonucleotide synthesis. *J. Biotechnol.* 94:15–22.

Nuwaysir, E. F., Huang, W., Albert, T. J., Singh, J., Nuwaysir, K., Pitas, A., Richmond, T., Gorski, T., Berg, J. P., Ballin, J., McCormick, M., Norton, J., Pollock, T., Sumwalt, T., Butcher, L., Porter, D., Molla, M., Hall, C., Blattner, F., Sussman, M. R., Wallace, R. L., Cerrina, F., and Green, R. D. (2002). Gene expression analysis using oligonucleotide arrays produced by maskless photolithography. *Genome Res.* 12:1749–1755.

Inkjet Arrays

Hughes, T. R., Mao, M., Jones, A. R., Burchard, J., Marton, M. J., Shannon, K. W., Lefkowitz, S. M., Ziman, M., Schelter, J. M., Meyer, M. R., Kobayashi, S., Davis, C., Dai, H., He, Y. D., Stephaniants, S. B., Cavet, G., Walker, W. L., West, A., Coffey, E., Shoemaker, D. D., Stoughton, R., Blanchard, A. P., Friend, S. H., and Linsley, P. S. (2001). Expression profiling using microarrays fabricated by an ink-jet oligonucleotide synthesizer. *Nature Biotechnol.* 19:342–347.

van't Veer, L. J., Dai, H., van de Vijver, M. J., He, Y. D., Hart, A. A., Mao, M., Peterse, H. L., van der Kooy, K., Marton, M. J., Witteveen, A. T., Schreiber, G. J., Kerkhoven, R. M., Roberts, C., Linsley, P. S., Bernards, R., and Friend, S. H. (2002). Gene expression profiling predicts clinical outcome of breast cancer. *Nature* 415:530–536.

Parallel Signature Sequencing

Brenner, S., Johnson, M., Bridgham, J., Golda, G., Lloyd, D. H., Johnson, D., Luo, S., McCurdy, S., Foy, M., Ewan, M., Roth, R., George, D., Eletr, S., Albrecht, G., Vermaas, E., Williams, S. R., Moon, K., Burcham, T., Pallas, M., DuBridge, R. B., Kirchner, J., Fearon, K., Mao, J., and Corcoran, K. (2000). Gene expression analysis by massively parallel signature sequencing (MPSS) on microbead arrays. *Nature Biotechnol.* 18:630–634.

Emerging Technologies

Marie, R., Jensenius, H., Thaysen, J., Christensen, C. B., and Boisen, A. (2002) . Adsorption kinetics and mechanical properties of thiol-modified DNA-oligos on gold investigated by microcantilever sensors. *Ultramicroscopy* 91:29–36.

2

Image Analysis

Image analysis is an important aspect of microarray experiments. It can have a potentially large impact on subsequent analysis such as clustering or the identification of differentially expressed genes.

—Yang et al. 2001a

Analysis of the image of the scanned array seeks to extract an intensity for each spot or feature on the array. In the simplest case, we seek one expression number for each gene. The analysis can be divided into several steps (Yang et al., 2001b):

1. Gridding
2. Segmentation
3. Intensity extraction
4. Background correction

This chapter first describes the basic concepts of image analysis and then lists a number of software packages available for the purpose.

2.1 GRIDDING

Whether you have a scanned image of a spotted array or an image of an Affymetrix GeneChip, you need to identify each spot or feature. That is accomplished by aligning a grid to the spots, because the spots are arranged in a grid of columns and rows (Figure 2.1). For photolithographically produced chips this may be easier than for robot spotted arrays, where more variation in the grid is possible. For the latter a manual intervention may be necessary to make sure that all the spots have been correctly identified.

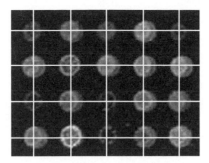

Figure 2.1   *Aligning a grid to identify the location of each spot.*

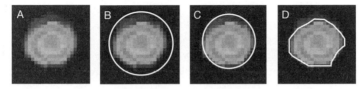

Figure 2.2   *Illustration of segmentation methods: (A) image before segmentation, (B) fixed circle segmentation, (C) adaptive circle segmentation, (D) adaptive shape segmentation.*

2.2   SEGMENTATION

Once the spots have been identified, they need to be separated from the background. The shape of each spot has to be identified. The simplest assumption is that all spots are circular of constant diameter. Everything inside the circle is assumed to be signal and everything outside is assumed to be background. This simple assumption rarely holds, and therefore most image analysis software includes some more advanced segmentation method. Adaptive circle segmentation estimates the diameter separately for each spot. Adaptive shape segmentation does not assume circular shape of each spot and instead tries to find the best shape to describe the spot. Finally, the histogram method analyzes the distribution of pixel intensities in and around each spot to determine which pixels belong to the spot and which pixels belong to the background (Figure 2.2).

2.3   INTENSITY EXTRACTION

Once the spot has been separated from the surrounding background, an intensity has to be extracted for each spot and potentially for each surrounding background. Typical measures are the mean or median intensity of all pixels within the spot.

2.4   BACKGROUND CORRECTION

On some array images, a slight signal is seen in the area that is in between spots. This is a background signal and it can be subtracted from the spot intensity to get a more accurate estimate of the biological signal from the spot. There are some problems associated with such a background correction, however. First, it does not necessarily follow that the background signal is added to the spot signal. In other words, some

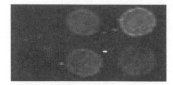

Figure 2.3 *Illustration of "ghost" where the background has higher intensity than the spot. In this case, subtracting background from spot intensity may be a mistake (see color insert).*

spots can be seen with lower intensity than the surrounding background. In that case, the surrounding background should clearly not be subtracted from the spot (Figure 2.3).

Second, a local estimation of background is necessarily associated with some noise. Subtracting such a noisy signal from a weak spot signal with noise will result in a number with even more noise. For weakly expressed genes this noise increase can negatively affect the following statistical analysis for differential expression.

The effect of not subtracting a background is that the absolute values may be slightly higher and that fold changes may be underestimated slightly. On balance, we choose not to subtract any local background but we do subtract a globally estimated background. This can, for example, be the second or third percentile of all the spot values. This is similar to the approach used by Affymetrix GeneChip software, where the image is segmented into 16 squares, and the average of the lower 2% of feature intensities for each block is used to calculate background. This background intensity is subtracted from all features within a block.

That leaves the issue of spatial bias on an array. This topic is usually considered under normalization. We have investigated spatial bias both for spotted arrays and Affymetrix arrays and found it to be significant in spotted arrays (Workman et al., 2002). We define spatial bias as an overall trend of fold changes that vary with the location on the surface (see Figure 2.4). Such a bias can be removed with Gaussian

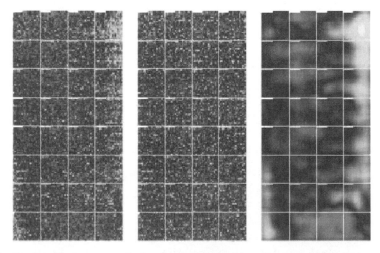

Figure 2.4 *Spatial effects on a spotted array. The blue-yellow color scale (see color insert) indicates fold change between the two channels (see color insert). A spatial bias is visible (left). Gaussian smoothing captures the bias (right), which can then be removed by subtraction from the image (center). (From Workman et al., 2002.)*

smoothing (Workman et al., 2002). In essence, the local bias in fold change is calculated in a window and subtracted from the observed fold change.

2.5 SOFTWARE

2.5.1 Free Software for Array Image Analysis

(From http://ihome.cuhk.edu.hk/~b400559/array.htm)

- *Dapple*
  Washington University
  (http://www.cs.wustl.edu/~jbuhler/research/dapple/)
- *F-Scan*
  National Institutes of Health
  (http://abs.cit.nih.gov/fscan/)
- *GridGrinder*
  Corning Inc.
  (http://gridgrinder.sourceforge.net/)
- *Matarray*
  Medical College of Wisconsin
  (http://www.mcw.edu/display/router.asp?docid=530)
- *P-Scan*
  National Institutes of Health
  (http://abs.cit.nih.gov/pscan/index.html)
- *ScanAlyze*
  Lawrence Berkeley National Laboratory
  (http://rana.lbl.gov/EisenSoftware.htm)
- *Spotfinder*
  The Institute for Genomic Research
  (http://www.tigr.org/software/tm4/spotfinder.html)
- *UCSF Spot*
  University of California, San Francisco
  (http://jainlab.ucsf.edu/Downloads.html)

2.5.2 Commercial Software for Array Image Analysis

(From http://ihome.cuhk.edu.hk/~b400559/array.htm)

- *AIDA Array Metrix*
  Raytest GmbH
  (http://www.raytest.de)
- *ArrayFox*
  Imaxia Corp.
  (http://www.imaxia.com/products.htm)

- *ArrayPro*
  Media Cybernetics, Inc.
  (http://www.mediacy.com/arraypro.htm)
- *ArrayVision*
  Imaging Research Inc.
  (http://www.imagingresearch.com/products/ARV.asp)
- *GenePix Pro*
  Axon Instruments, Inc.
  (http://www.axon.com/GN_GenePixSoftware.html)
- *ImaGene*
  BioDiscovery, Inc.
  (http://www.biodiscovery.com/imagene.asp)
- *IconoClust*
  CLONDIAG Chip Technologies GmbH
  (http://www.clondiag.com/products/sw/iconoclust/)
- *Microarray Suite*
  Scanalytics, Inc.
  (http://www.scanalytics.com/product/microarray/index.shtml)
- *Koadarray*
  Koada Technology
  (http://www.koada.com/koadarray/)
- *Lucidea*
  Amersham Biosciences
  (http://www1.amershambiosciences.com/)
- *MicroVigene*
  VigeneTech, Inc.
  (http://www.vigenetech.com/product.htm)
- *Phoretics Array*
  Nonlinear Dynamics
  (http://www.phoretix.com/products/array_products.htm)
- *Quantarray*
  PerkinElmer, Inc.
  (http://las.perkinelmer.com/)
- *Spot*
  CSIRO Mathematical and Information Sciences
  (http://experimental.act.cmis.csiro.au/Spot/index.php)

2.6 SUMMARY

The software that comes with the scanner is usually a good start for image analysis. It is better to use a global background correction instead of subtracting a locally estimated background from each spot. It is always a good idea to look at the image of a chip to observe any visible defects, bubbles, or clear spatial bias.

For all the research into how best to analyze microarray data, the image analysis field has been largely ignored. We are still doing image analysis in much the same way as we did when the field emerged less than ten years ago. It is obvious that further research will lead to improved image analysis that will lead to better accuracy in medical diagnostics.

3

Basic Data Analysis

In gene expression analysis, technological problems and biological variation make it difficult to distinguish signal from noise. Once we obtain reliable data, we look for patterns and need to determine their significance.

—Vingron, 2001

3.1 NORMALIZATION

Microarrays are usually applied to the comparison of gene expression profiles under different conditions. That is because most of the biases and limitations that affect absolute measurements do not affect relative comparisons. There are a few exceptions to that. One is that you have to make sure that what you are comparing is really comparable. The chips have to be the same under the different conditions, but also the amount of sample applied to each chip has to be comparable. An example will illustrate this. Figure 3.1 shows a comparison between two chips where the same labeled RNA has been added to both. Ideally, all the intensity measurements on one chip should match those on the other, all points should lie on the diagonal. They do not, and they reveal both random and systematic bias. The systematic bias is revealed by a deviation from the diagonal that increases with intensity. This is a systematic bias that is signal dependent. It becomes even more pronounced when we plot the logarithm of the ratio versus the logarithm of the intensity (Figure 3.1B). This is often referred to as an M vs. A plot or MVA plot and it is often used to identify signal-dependent biases.

We cannot remove the random bias (we will deal with it later by using replicates), but we can remove systematic bias. This is often referred to as normalization. Normalization is based on some assumptions that identify reference points.

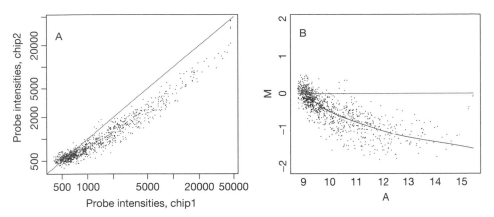

Figure 3.1 *Comparison of the probe intensities between two Affymetrix chips with the same labeled RNA applied. (A) Log of intensity versus log of intensity. (B) Log of ratio (M = log(chip1/chip2)) versus average log intensity (A = (log chip1 + log chip2)/2). The curve in (B) shows a lowess (locally weighted least squares fit) applied to the data.*

3.1.1 One or More Genes Assumed Expressed at Constant Rate

These genes are referred to as housekeeping control genes. Examples include the *GAPDH* gene. Multiply all intensities by a constant until the expression of the control gene is equal in the arrays that are being compared. For arrays with few genes this is often the only normalization method available. This is, however, a linear normalization that does not remove the observed signal-dependent nonlinearity. In the MVA plot of Figure 3.1B it amounts to addition of a constant to the ratio to yield the normalized data of Figure 3.2A. The systematic bias is still present. It would lead you to conclude that weakly expressed genes are upregulated on chip 1 relative to chip 2, whereas highly expressed genes are downregulated on chip 1 relative to chip 2. This conclusion we know is false, because we put the same RNA on both chips. If you have more control genes with different intensity you can draw a normalization curve by fitting a curve through them. That makes the normalization signal dependent, similar to what is seen in Figure 3.2B.

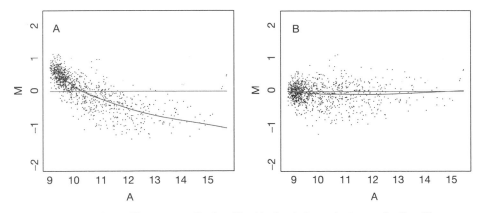

Figure 3.2 *Comparison of linear normalization (A) with signal-dependent normalization (B).*

3.1.2   Sum of Genes Is Assumed Constant

This assumes that the total messenger RNA in a cell is constant, which may hold true if you are using a large transcriptome chip with thousands of genes. It is the approach used by Affymetrix in the GeneChip software. All intensities on one chip are multiplied by a constant until they have the same sum or average as the chip you want to compare to. The problem is that this is still a linear normalization that does not remove the signal-dependent bias (Figure 3.2A). If your array contains only a few hundred genes that were selected for their participation in a certain process, the assumption of constant sum may not be a good one. Then it may be better to rely on one or more housekeeping control genes.

3.1.3   Subset of Genes Is Assumed Constant

Li and Wong (2001b) have proposed a method whereby the constant control genes are not known a priori but are instead identified as genes whose intensity rank does not differ more than a threshold value between two arrays. This invariant set is defined iteratively and used to draw a normalization curve that is signal dependent. In practice, this method works extremely well, and it has compared well to all other normalization methods developed.

3.1.4   Majority of Genes Assumed Constant

We have developed a signal-dependent normalization method, qspline (Workman et al., 2002), that assumes that the overall intensity distributions between two arrays should be comparable. That means that the quantiles[1] of the distributions, plotted in a quantile–quantile plot, should lie on the diagonal. If they do not, they form a normalization curve that is signal dependent. After normalization, the scatter point smoothing line is close to a straight line through zero in the MVA plot (Figure 3.2B). We have compared our method, qspline, to the methods of Li and Wong (2001b) and Irizarry et al. (2003b) and found them to behave very similarly on the datasets we have tested. For spotted arrays, we have found our qspline method to perform slightly better on one dataset than lowess as proposed by Yang et al. (2002).

Even if the assumption does not hold, and less than a majority of genes are constant, the normalization still works provided that the number of upregulated genes roughly equals the number of downregulated genes and provided that there is no signal-dependent bias in up- or downregulation.

3.1.5   Logit Normalization

Several years after our signal-dependent normalization method and similar methods were published, an entirely new approach to normalization was published (Lemon et al., 2003). In light of its simplicity, the performance is surprisingly good.

$$\text{logit}(y) = \log\left(\frac{y - \text{background}}{\text{saturation} - y}\right),$$

---

[1]A quantile has a fixed number of genes below it in intensity. The first quantile could have 1% of the genes below it in intensity, the second quantile 2% of the genes below it in intensity, and so on.

where background is calculated as the minimum intensity measured on the chip minus 0.1% of the signal intensity range: $min -0.001 * (max - min)$; and saturation is calculated as the maximum intensity measured on the chip plus 0.1% of the signal intensity range: $max +0.001 * (max - min)$. The result is then Z-transformed to mean zero and standard deviation 1. The logit normalization is usually performed together with a median P-value statistical analysis and the combination has, in our hands, led to improvements in classification accuracy compared to other normalization methods.

3.1.6 Spike Controls

If none of the above assumptions seem applicable to your experiment, there is one choice of last resort. You can add a spiked control to your mRNA preparation. The idea is to measure the amount of mRNA or total RNA extracted from the cell, and then add a known transcript of known concentration to the pool. This spiked transcript is then assumed to be amplified and labeled the same way as the other transcripts and detected with a unique probe on the array. The spiked transcript must not match any gene in your RNA preparation, so for a human RNA preparation an *Escherichia coli* gene could be used. After scanning the array you multiply all measurements on one array until the spiked control matches that on the other array.

This approach has the limitation that it results in a linear normalization that does not correct signal-dependent bias (unless you use many spiked control genes with different concentration). Finally, it is only as accurate as the accuracy of measuring the total amount of RNA and the accuracy of adding an exact amount of spiked transcript.

3.2 DYE BIAS, SPATIAL BIAS, PRINT TIP BIAS

The above mentioned global normalization methods (in particular, the signal-dependent ones) work well for factory-produced oligonucleotide arrays such as Affymetrix GeneChips. For arrays spotted with a robot, however, there may be substantial residual bias after a global normalization. First, there is a dye bias. The dyes Cy3 and Cy5 have different properties that are revealed if you label the same sample with both dyes and then plot the resulting intensities against each other. Such a plot reveals that the bias is signal dependent. For that reason, the signal-dependent normalization methods mentioned above will also remove the dye bias, and after signal-dependent normalization you can directly compare Cy3 channels to Cy5 channels. If you are unable to use a signal-dependent normalization method, a typical approach to removing dye bias is to use a dye swap, which means that you label each sample with both Cy3 and Cy5 and then take the ratio of the averages of each sample. Note, however, that dye swap normalization does not remove signal-dependent bias beyond the dye bias.

Spatial bias was dealt with in Chapter 2. Removing it requires knowledge of the layout of the array.

Finally, it is often possible to observe a print tip bias. The spots on the array are usually not printed with the same printing tip. Instead, several tips are used to print in parallel. So the spots on the array are divided into groups, where each group of spots have been made with the same tip. When you compare the average log ratio within the print tip groups, you can observe differences between print tip groups. These can be due to a nonrandom order in which genes are printed, or they can be due to spatial biases, or they can be due to true physical differences between the print tips. To remove the print tip bias is relatively straightforward. First you perform a linear normalization of each print tip group until their average signal ratio is equal. Then you perform a global signal-dependent normalization to remove any signal-dependent bias.

3.3 EXPRESSION INDICES

For spotted arrays using only one probe for each gene, you can calculate fold changes after normalization. For Affymetrix GeneChips and other technologies relying on several different probes for each gene, it is necessary to condense these probes into a single intensity for each gene. This we refer to as an expression index.

3.3.1 Average Difference

Affymetrix, in the early version (MAS 4.0) of their software, calculated an Average Difference between probe pairs. A probe pair consists of a perfect match (PM) oligo and a mismatch (MM) oligo for comparison. The mismatch oligo differs from the perfect match oligo in only one position and is used to detect nonspecific hybridization. Average Difference was calculated as follows:

$$\text{AvgDiff} = \frac{\sum_N (\text{PM} - \text{MM})}{N},$$

where N is the number of probe pairs used for the calculation (probe pairs that deviate by more than 3 standard deviations from the mean are *excluded* from the calculation). If the AvgDiff number is negative or very small, it means that either the target is absent or there is nonspecific hybridization. Affymetrix calculates an Absolute Call based on probe statistics: Absent, Marginal, or Present (refer to the Affymetrix manual for the decision matrix used for making the Absolute Call).

3.3.2 Signal

In a later version of their software (MAS 5.0), Affymetrix has replaced AvgDiff with a Signal, which is calculated as

$$\text{Signal} = \text{Tukeybiweight}[\log(\text{PM}_n - \text{CT}_n)],$$

where Tukeybiweight is a robust estimator of central tendency. To avoid negative numbers when subtracting the mismatch, a number CT is subtracted that can never be larger than PM. Note, however, that this could affect the normality assumption often used in downstream statistical analysis (Giles and Kipling, 2003).

3.3.3 Model-Based Expression Index

Li and Wong (2001a, b) instead calculate a weighted average difference:

$$\tilde{\theta} = \frac{\sum_N (\text{PM}_n - \text{MM}_n)\phi_n}{N},$$

where ϕ_n is a scaling factor that is specific to probe pair $PM_n - MM_n$ and is obtained by fitting a statistical model to a series of experiments. This model takes into account that probe pairs respond differently to changes in expression of a gene and that the variation between replicates is also probe-pair dependent. Li and Wong have also shown that the model works without the mismatches (MM) and usually has lower noise than when mismatches are included. Software for fitting the model (weighted average difference and weighted perfect match), as well as for detecting outliers and obtaining estimates on reliability, is available for download (http://www.dchip.org). Lemon and co-workers (2002) have compared the Li–Wong model to the Affymetrix Average Difference and found it to be superior in a realistic experimental setting. Note that model parameter estimation works best with 10 to 20 chips.

3.3.4   Robust Multiarray Average

Irizarry et al. (2003a) have published a Robust Multiarray Average that also reduces noise by omitting the information present in the mismatch probes of Affymetrix GeneChips:

$$RMA = \text{Medianpolish}[\log PM_n - \alpha_n)],$$

where Medianpolish is a robust estimator of central tendency and α is a scaling factor that is specific to probe PM_n and is obtained by fitting a statistical model to a series of experiments.

3.3.5   Position-Dependent Nearest Neighbor Model

All the above expression indices are a statistical treatment of the probe data that assume that the performance of a probe can be estimated from the data. While this is true, there may be an even more reliable way of estimating probe performance: based on thermodynamics. The field of probe design for microarrays has been hampered by an absence of good thermodynamic models that accurately describe hybridization to an oligo attached to an array surface. Zhang et al. (2003a,b) have developed just that. They extend the nearest neighbor energy model, which works well for oligonucleotides in solution, with a position term that takes into account whether a nucleotide (or pair of nucleotides) is at the center of a probe or near the array surface or near the free end of the probe. They can use real array data to estimate the parameters of this model, and the resulting model works quite well at modeling the sequence-dependent performance of each probe. As such, it can be used for condensing the individual probe measures into one number for each gene.

3.3.6   Logit-t

Lemon et al. (2003) took a radically different approach. Instead of condensing the individual probes of a probeset into one number, they perform a t-test for each PM probe in the probeset. Then they use the median *P-value* of all PM probes to determine whether the gene is differentially expressed. In our hands, this approach has been superior in selecting genes for classification in medical diagnostics.

3.4 DETECTION OF OUTLIERS

Outliers in chip experiments can occur at several levels. You can have an entire chip that is bad and consistently deviates from other chips made from the same condition or sample. Or you can have an individual gene on a chip that deviates from the same gene on other chips from the same sample. That can be caused by image artifacts such as hairs, air bubbles, precipitation, and so on. Finally, it is possible that a single probe, due to precipitation or other artifact, is perturbed.

How can you detect outliers in order to remove them? Basically, you need a statistical model of your data. The simplest model is equality among replicates. If one replicate (chip, gene, or probe) deviates several standard deviations from the mean, you can consider it an outlier and remove it. The *t*-test measures standard deviation and gives genes where outliers are present among replicates a low significance (see Section 3.6).

More advanced statistical models have been developed that also allow for outlier detection and removal (Li and Wong, 2001a,b).

3.5 FOLD CHANGE

Having performed normalization (and, if necessary, expression index condensation), you should now be able to compare the expression level of any gene in the sample to the expression level of the same gene in the control. The next thing you want to know is: How many fold up- or downregulated is the gene, or is it unchanged?

The simplest approach to calculate fold change is to divide the expression level of a gene in the sample by the expression level of the same gene in the control. Then you get the fold change, which is 1 for an unchanged expression, less than 1 for a downregulated gene, and larger than 1 for an upregulated gene. The definition of fold change will not make any sense if the expression value in the sample or in the control is zero or negative. Early Average Difference values from Affymetrix sometimes were, and a quick-and-dirty way out of this problem was to set all Average Difference values below 20 to 20. This was the approach usually applied.

The problem with fold change emerges when one takes a look at a scale. Upregulated genes occupy the scale from 1 to infinity (or at least 1000 for a 1000-fold upregulated gene) whereas all downregulated genes only occupy the scale from 0 (0.001 for a 1000-fold downregulated gene) to 1. This scale is highly asymmetric.

The Affymetrix GeneChip software (early version MAS 4.0) calculates fold change in a slightly different way, which does stretch out that scale to be symmetric:

$$\text{AffyFold} = \frac{\text{Sample} - \text{Control}}{\min(\text{Sample}, \text{Control})} + \begin{cases} +1 & \text{if Sample} > \text{Control} \\ -1 & \text{if Sample} < \text{Control} \end{cases},$$

where Sample and Control are the AvgDiffs of the sample and the control, respectively. For calculation of fold change close to the background level, consult the Affymetrix manual.

This function is discontinuous and has no values in the interval from −1 to 1. Upregulated genes have a fold change greater than 1 and downregulated genes have a fold change less than −1. But the scale for downregulated genes is comparable to the scale for upregulated genes.

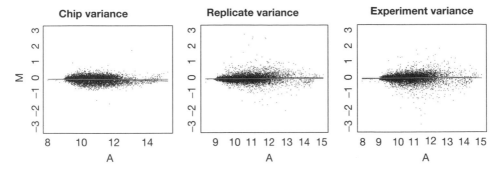

Figure 3.3 *Observed fold changes when comparing (left) chips with the same labeled mRNA, (middle) chips with mRNA preparations from two replicate cultures, and (right) chips from two different experimental conditions. Plots show log of ratio (M = log(chip1/chip2)) versus average log intensity (A = (log chip1 + log chip2)/2) for all genes.*

Both the fold change and Affyfold expressions are intuitively rather easy to grasp and deal with, but for further computational data analysis they are not useful, either because they are asymmetric or because they are discontinuous. For further data analysis you need to calculate the logarithm of fold change. Logfold, as we will abbreviate it, is undefined for the Affymetrix fold change but can be applied to the simple fold change provided that you have taken the precaution to avoid values with zero or negative expression.

It is not important whether you use the natural logarithm ($\log_e$), base 2 logarithm ($\log_2$), or base 10 logarithm ($\log_{10}$).

3.6 SIGNIFICANCE

If you have found a gene that is twofold upregulated ($\log_{10}$ fold is 0.3) from a normal subject to a diseased subject, how do you know whether this is not just a result of experimental error? You need to determine whether or not a twofold regulation is *significant* (see Figure 3.3). There are many ways to estimate significance in chip experiments, but most are based on replication. You must have a group of normal subjects and a group of diseased subjects, and look for those changes that are consistent. If there are only two groups, the *t*-test can be used to determine whether the expression of a particular gene is significantly different between the two groups.

The *t*-test looks at the mean and variance of the sample and control distributions and calculates the probability that the observed difference in mean occurs when the null hypothesis is true[2].

When using the *t*-test it is often assumed that there is equal variance between sample and control. That allows the sample and control to be pooled for variance estimation. If the variance cannot be assumed equal, you can use Welch's *t*-test, which assumes unequal variances of the two populations.

When using the *t*-test for analysis of microarray data, it is often a problem that the number of replicates is low. The lower the number of replicates, the more difficult it

---

[2]The null hypothesis states that the mean of the two distributions is equal. Hypothesis testing allows us to calculate the probability of finding the observed data when the hypothesis is true. To calculate the probability we make use of the normal distribution. When the probability is low, we reject the null hypothesis.

will be to estimate the variance. For only two replicates it becomes almost impossible. There are several solutions to this. The simplest solution is to take fold change into account for experiments with low number of replicates (say, less than 3), and not consider genes that have less than two fold change in expression. This will guard against low P-values that arise from underestimation of variance and is similar to the approach used in SAM, where a constant is added to the gene-specific variance (Tusher et al., 2001). The other possibility is to base variance estimation not only on a single gene measurement, but to include variance estimates from the whole population. Several methods have been developed for this purpose (Baldi and Long, 2001; Kerr and Churchill, 2001a; Lonnstedt and Speed, 2002).

3.6.1 Multiple Conditions

If there are more than two conditions, the t-test may not be the method of choice, because the number of comparisons grow if you perform all possible comparisons between conditions. The method analysis of variance (ANOVA) will, using the F distribution, calculate the probability of finding the observed differences in means between more than two conditions when the null hypothesis is true (when there is no difference in means).

Software for running the t-test and ANOVA will be discussed in Section 10.5. A web-based method for t-test is available (http://visitor.ics.uci.edu/genex/cybert/) (Baldi and Long, 2001). Baldi and Long (2001) recommend using the t-test on log-transformed data.

3.6.2 Nonparametric Tests

Both the t-test and ANOVA assume that your data follow the normal distribution. Although both methods are robust to moderate deviations from the normal distribution[3], alternative methods exist for assessing significance without assuming normality. The Wilcoxon/Mann-Whitney rank sum test will do the same without using the actual expression values from the experiment, only their rank relative to each other.

When you rank all expression levels from the two conditions, the best separation you can have is that all values from one condition rank higher than all values from the other condition. This corresponds to two nonoverlapping distributions in parametric tests. But since the Wilcoxon test does not measure variance, the significance of this result is limited by the number of replicates in the two conditions. It is for this reason that you may find that the Wilcoxon test for low numbers of replication gives you a poor significance and that the distribution of P-values is rather granular.

3.6.3 Correction for Multiple Testing

For all statistical tests that calculate a P-value, it is important to consider the effect of multiple testing as we are looking at not just one gene but thousands of genes. If a P-value of 0.05 tells you that you have a probability of 5% of making a type I error (false positive) on one gene, then you expect 500 type I errors (false positive genes) if you look at 10,000 genes. That is usually not acceptable. You can use the Bonferroni correction to reduce the significance cutoff to a level where you again have

---

[3]Giles and Kipling (2003) have demonstrated that deviations from the normal distribution are small for most microarray data, except when using the Affymetrix MAS 5.0 software.

only 5% probability of making one or more type I errors among all 10,000 genes. This new cutoff is $0.05/10,000 = 5 \times 10^{-6}$. That is a pretty strict cutoff, and for many purposes you can live with more type I errors than that. Say you end up with a list of 100 significant genes and you are willing to accept 5 type I errors (false positives) on this list. Then you are willing to accept a *False Discovery Rate (FDR)* of 5%. Benjamini and Hochberg (1995) have come up with a method for controlling the FDR at a specified level. After ranking the genes according to significance (*P*-value) and starting at the top of the list, you accept all genes where

$$P \leq \frac{i}{m} q,$$

where i is the number of genes accepted so far, m is the total number of genes tested, and q is the desired FDR. For $i > 1$ this correction is less strict than a Bonferroni correction.

The False Discovery Rate can also be assessed by permutation. If you permute the measurements from sample and control and repeat the *t*-test for all genes, you get an estimate of the number of type I errors (false positives) that can be expected at the chosen cutoff in significance. When you divide this number by the number of genes that pass the *t*-test on the unpermuted data, you get the FDR. This is the approach used in the software SAM (Tusher et al., 2001).

If no genes in your experiment pass the Bonferroni or Benjamini–Hochberg corrections, then you can look at those that have the smallest *P*-value. When you multiply their *P*-value by the number of genes in your experiment, you get an estimate of the number of false positives. Take this false positive rate into account when planning further experiments.

3.6.4   Example I: *t*-Test and ANOVA

A small example using only four genes will illustrate the *t*-test and ANOVA. The four genes are each measured in six patients, which fall into three categories: normal (N), disease stage A, and disease stage B. That means that each category has been *replicated* once (Table 3.1).

We can perform a *t*-test (see Section 10.5 for details) to see if genes differ significantly between patient category A and patient category B (Table 3.2). But you should be careful performing a *t*-test on as little as two replicates in real life. This is just for illustration purposes.

Gene b is significantly different at a 0.05 level, even after multiplying the *P*-value by 4 to correct for multiple testing. Gene a is not significant at a 0.05 level after

TABLE 3.1   Expression Readings of Four Genes in Six Patients

| Gene | Patient | | | | | |
	N_1	N_2	A_1	A_2	B_1	B_2
a	90	110	190	210	290	310
b	190	210	390	410	590	610
c	90	110	110	90	120	80
d	200	100	400	90	600	200

TABLE 3.2 *t*-Test on Difference Between Patient Categories A and B

Gene	Patient				P-Value
	A_1	A_2	B_1	B_2	
a	190	210	290	310	0.019
b	390	410	590	610	0.005
c	110	90	120	80	1.000
d	400	90	600	200	0.606

TABLE 3.3 ANOVA on Difference Between Patient Categories N, A, and B

Gene	Patient						P-Value
	N_1	N_2	A_1	A_2	B_1	B_2	
a	90	110	190	210	290	310	0.0018
b	190	210	390	410	590	610	0.0002
c	90	110	110	90	120	80	1.0000
d	200	100	400	90	600	200	0.5560

Bonferroni correction, and genes *c* and *d* have a high probability of being unchanged. For gene *d* that is because, even though an increasing trend is observed, the variation within each category is too high to allow any conclusions.

If we perform an ANOVA instead, testing for genes that are significantly different in at least one of three categories, the picture changes slightly (Table 3.3).

In the ANOVA, both genes *a* and *b* are significant at a 0.01 level even after Bonferroni correction. So taking all three categories into account increases the power of the test relative to the *t*-test on just two categories.

3.6.5 Example II: Number of Replicates

If replication is required to determine the significance of results, how many replicates are required? An example will illustrate the effect of the number of replicates. We have performed six replicates of each of two conditions in a *Saccharomyces cerevisiae* GeneChip experiment (Piper et al., 2002). Some of the replicates have even been performed in different labs. Assuming the results of a *t*-test on this dataset to be the correct answer, we can ask: How close would five replicates have come to that answer? How close would four replicates have come to that answer? We have performed this test (Piper et al., 2002) and Table 3.4 shows the results. For each choice of replicates, we show how many false positives (Type I errors) we have relative to the correct answer and how many false negatives (Type II errors) we have relative to the correct answer. The number of false positives in the table is close to the number we have chosen with our cutoff in the Bonferroni corrected *t*-test (a 0.005 cutoff at 6383 genes yields 32 expected false positives). The number of false negatives, however, is greatly affected by the number of replicates.

Table 3.5 shows that the *t*-test performs almost as well as SAM[4] (Tusher et al., 2001), which has been developed specifically for estimating the false positive rate in DNA microarray experiments based on permutations of the data.

---

[4]Software is available for download at http://www-stat.stanford.edu/~tibs/SAM/index.html.

TABLE 3.4   Effect of Number of Replicates on Type I (FP) and II (FN) Errors in *t*-Test

	Number of Replicates of Each Condition				
	2	3	4	5	6
True positives	23	144	405	735	1058
False positives	8	18	29	45	0
False negatives	1035	914	653	323	0

TABLE 3.5   Effect of Number of Replicates on Type I and II Errors in SAM (Tusher et al., 2001)

	Number of Replicates of Each Condition				
	2	3	4	5	6
True positives	27	165	428	748	1098
False positives	3	4	14	27	0
False negatives	1071	933	670	350	0

While this experiment may not be representative, it does illustrate two important points about the *t*-test. You can control the number of false positives even with very low numbers of replication. But you lose control over the false negatives as the number of replications goes down.

So how many replicates do you have to perform to avoid any false negatives? That depends mainly on two parameters: how large the variance is between replicates and how small a fold change you wish to detect. Given those, it is possible to calculate the number of replicates needed to achieve a certain power (1 minus the false negative rate) in the *t*-test[5].

3.7   MIXED CELL POPULATIONS

The analysis presented above assumes that we are looking at pure cell populations. In cell cultures we are assuming that all cells are identical, and in tissue samples we are assuming that the tissue is homogeneous. The degree to which these assumptions are true varies from experiment to experiment. If you have isolated a specific cell type from blood, there may still be a mixture of subtypes within this population. A tissue sample may contain a combination of tumor and normal cells. A growing cell culture contains cells in different stages of the cell cycle.

If the proportion of subtypes is constant throughout the experiment, a standard analysis can be applied. You just have to remember that any signal arising from a single subtype will be diluted by the presence of other subtypes.

If the proportion of subtypes varies in the experiment, it may be possible to resolve mathematically the proportions and estimate the expression of individual subtypes within the population. But the mathematical procedures for doing this again rest on a number of assumptions. If you assume that each cell subtype has a uniform and unchanging expression profile, and that the only thing that changes in your experiment is the ratio between cell subtypes, the problem becomes a simple mathematical problem

---

[5]For example, by using the `power.t.test` function of the R package available from www.r-project.org

of solving linear equations (Lu et al., 2003). First you need to obtain the expression profile of each of the pure cell subtypes in an isolated experiment. Then you find the linear combination of the pure profile that best fits the data of the mixed cell population experiments. This gives the proportions of the individual cell subtypes.

Another way of separating the samples into their constituent cell types is if you have more samples than cell types. Then the problem may become determined under a number of constraints (Venet et al., 2001):

$$\mathbf{M} = \mathbf{GC},$$

where $\mathbf{M}$ is the matrix of measured values (rows correspond to genes and columns correspond to experiments), $\mathbf{G}$ is the expression profile of each cell type (rows correspond to genes and columns correspond to cell types), and $\mathbf{C}$ is the concentration matrix (rows correspond to cell types and columns correspond to measurements). Under a number of assumptions and constraints, it may be possible to find an optimal solution $\mathbf{G}$ and $\mathbf{C}$ from $\mathbf{M}$ (Venet et al., 2001).

3.8 SUMMARY

Whether you have intensities from a spotted array or Signal (use the Li–Wong weighted PM, if possible) from an Affymetrix chip, the following suggestions apply:

- The standard normalization with one factor to get the same average intensity in all chips is a good way to start, but it is not the best way to do it. Use signal-dependent normalization or logit normalization if possible.
- Repeat each condition of the experiment (as a rule-of-thumb at least three times) and apply a statistical test for significance of observed differences. Apply the test on the normalized intensities (or expression indices). For spotted arrays with large variation between slides, you can consider applying the statistical test on the fold change from each slide as well.
- Correct the statistical test for multiple testing (Bonferroni correction or similar).

Software for all the methods described in this chapter is available from www.bioconductor.org, which will be described in more detail in Chapter 10.

FURTHER READING

Normalization

Bolstad, B. M., Irizarry, R. A., Astrand, M., and Speed, T. P. (2003). A comparison of normalization methods for high density oligonucleotide array data based on variance and bias. *Bioinformatics* 19(2):185–193.

Dudoit, S., Yang, Y., Callow, M. J., and Speed, T. P. (2000). Statistical methods for identifying differentially expressed genes in replicated cDNA microarray experiments. Technical report #578, August 2000. (Available at http://www.stat.berkeley.edu/tech-reports/index.html.)

Goryachev, A. B., Macgregor, P. F., and Edwards, A. M. (2001). Unfolding of microarray data. *J. Comput. Biol.* 8:443–461.

Schadt, E. E., Li, C., Su, C., and Wong, W. H. (2000). Analyzing high-density oligonucleotide gene expression array data. *J. Cell. BioChem.* 80:192–201.

Schuchhardt, J., Beule, D., Malik, A., Wolski, E., Eickhoff, H., Lehrach, H., and Herzel, H. (2000). Normalization strategies for cDNA microarrays. *Nucleic Acids Res.* 28:e47.

Zien, A., Aigner, T., Zimmer, R., and Lengauer, T. (2001). Centralization: A new method for the normalization of gene expression data. *Bioinformatics* 17(Suppl 1):S323–S331.

Expression Index Calculation

Lazaridis, E. N., Sinibaldi, D., Bloom, G., Mane, S., and Jove, R. (2002). A simple method to improve probe set estimates from oligonucleotide arrays. *Math. Biosci.* 176(1):53–58.

Nonparametric Significance Tests Developed for Array Data

Efron, B., and Tibshirani, R. (2002). Empirical Bayes methods and false discovery rates for microarrays. *Genet. Epidemiol.* 23(1):70–86.

Park, P. J., Pagano, M., and Bonetti, M. (2001). A nonparametric scoring algorithm for identifying informative genes from microarray Data. *Pacific Symposium on Biocomputing* 6:52–63.(Manuscript available online at http://psb.stanford.edu.)

Student's *t*-test, ANOVA, and Wilcoxon/Mann-Whitney

Kerr, M. K., Martin, M., and Churchill, G. A. (2000). Analysis of variance for gene expression microarray data. *J. Comput. Biol.* 7:819–837.

Montgomery, D. C., and Runger, G. C. (1999). *Applied Statistics and Probability for Engineers.* New York: Wiley.

Number of Replicates

Black, M. A., and Doerge, R. W. (2002). Calculation of the minimum number of replicate spots required for detection of significant gene expression fold change in microarray experiments. *Bioinformatics* 18(12):1609–1616.

Lee, M. L., and Whitmore, G. A. (2002). Power and sample size for DNA microarray studies. *Stat. Med.* 21(23):3543–3570.

Pan, W., Lin, J., and Le, C. (2002). How many replicates of arrays are required to detect gene expression changes in microarray experiments? A mixture model approach. *Genome Biol.* 3(5):research0022.

Pavlidis, P., Li, Q., and Noble, W. S. (2003). The effect of replication on gene expression microarray experiments. *Bioinformatics* 19(13):1620–1627.

Wahde, M., Klus, G. T., Bittner, M. L., Chen, Y., and Szallasi, Z. (2002). Assessing the significance of consistently mis-regulated genes in cancer associated gene expression matrices. *Bioinformatics* 18(3):389–394.

Correction for Multiple Testing

Bender, R., and Lange, S. (2001). Adjusting for multiple testing—when and how? *J. Clin. Epidemiol.* 54:343–349.

Dudoit, S., Yang, Y., Callow, M. J., and Speed, T. P. (2000). Statistical methods for identifying differentially expressed genes in replicated cDNA microarray experiments. Technical report #578, August 2000. (Available at http://www.stat.berkeley.edu/tech-reports/index.html.)

Reiner, A., Yekutieli, D., and Benjamini, Y. (2003). Identifying differentially expressed genes using false discovery rate controlling procedures. *Bioinformatics* 19(3):368–375.

Variance Stabilization

Huber, W., Von Heydebreck, A., Sultmann, H., Poustka, A., and Vingron, M. (2002). Variance stabilization applied to microarray data calibration and to the quantification of differential expression. *Bioinformatics* 18 (Suppl 1):S96–S104.

Other Significance Tests Developed for Array Data

Baggerly, K. A., Coombes, K. R., Hess, K. R., Stivers, D. N, Abruzzo, L. V., and Zhang, W. (2001). Identifying differentially expressed genes in cDNA microarray experiments. *J. Comput. Biol.* 8(6):639–659.

Ideker, T., Thorsson, V., Siegel, A. F., and Hood, L. (2000). Testing for differentially-expressed genes by maximum-likelihood analysis of microarray data. *J. Comput. Biol.* 7:805–817.

Newton, M. A., Kendziorski, C. M., Richmond, C. S., Blattner, F. R., and Tsui, K. W. (2001). On differential variability of expression ratios: improving statistical inference about gene expression changes from microarray data. *J. Comput. Biol.* 8:37–52.

Rocke, D. M., and Durbin, B. (2001). A model for measurement error for gene expression arrays. *J. Comput. Biol.* 8(6):557–569.

Theilhaber, J., Bushnell, S., Jackson, A., and Fuchs, R. (2001). Bayesian estimation of fold-changes in the analysis of gene expression: the PFOLD algorithm. *J. Comput. Biol.* 8(6):585–614.

Thomas, J. G., Olson, J. M., Tapscott, S. J., and Zhao, L. P. (2001). An efficient and robust statistical modeling approach to discover differentially expressed genes using genomic expression profiles. *Genome Res.* 11:1227–1236.

Wolfinger, R. D., Gibson, G., Wolfinger, E. D., Bennett, L., Hamadeh, H., Bushel, P., Afshari, C., and Paules, R. S. (2001). Assessing gene significance from cDNA microarray expression data via mixed models. *J. Comput. Biol.* 8(6):625–637.

Zhao, L. P., Prentice, R., and Breeden, L. (2001). Statistical modeling of large microarray data sets to identify stimulus–response profiles. *Proc. Natl. Acad. Sci. USA* **98**:5631–5636.

4

Visualization by Reduction of Dimensionality

The data from expression arrays are of high dimensionality. If you have measured 6000 genes in 15 patients, the data constitute a matrix of 15 by 6000. It is impossible to discern any trends by visual inspection of such a matrix. It is necessary to reduce the dimensionality of this matrix to allow visual analysis. Since visual analysis is traditionally performed in two dimensions, in a coordinate system of x and y, many methods allow reduction of a matrix of any dimensionality to only two dimensions. These methods include principal component analysis, independent component analysis, correspondence analysis, multidimensional scaling, and cluster analysis.

4.1 PRINCIPAL COMPONENT ANALYSIS

If we want to display the data in just two dimensions, we want to capture as much of the variation in the data as possible in just these two dimensions. Principal component analysis (PCA) has been developed for this purpose. Imagine 6000 genes as points in a 15-dimensional hyperspace, each dimension corresponding to expression in one of 15 patients. You will see a cloud of 6000 points in hyperspace. But the cloud is not hyperspherical. There will be one direction in which the cloud will be more irregular or extended. (Figure 4.1 illustrates this with only a few points in three dimensions.) This is the direction of the first principal component. Mathematically, this direction is defined as the direction of the largest eigenvector of the covariance matrix of the genes. This direction will not necessarily coincide with one of the patient axes. Rather, it will have projections of several, or all, patient axes on it. Next, we look for a direction that is orthogonal to the first principal component

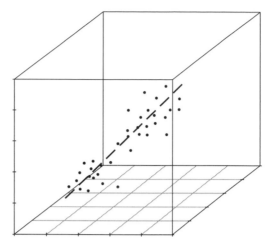

Figure 4.1 *A cloud of points in three-dimensional space. The cloud is not regular. It extends more in one direction than in all other directions. This direction is the first principal component (dashed line).*

and captures the maximum amount of variation left in the data. This is the second principal component. We can now plot all 6000 genes in these two dimensions. We have reduced the dimensionality from 15 to 2, while trying to capture as much variation in the data as possible. The two principal components have been constructed as sums of the individual patient axes.

What will this analysis tell you? Perhaps nothing; it depends on whether there is a trend in your data that is discernible in two dimensions. Other relationships can be visualized with cluster analysis, which will be described in Chapter 5.

Instead of reducing the patient dimensions we can reduce the gene dimensions. Why not throw out all those genes that show no variation? We can achieve this by performing a principal component analysis of the genes. We are now imagining our data as 15 points in a space of 6000 dimensions, where each dimension records the expression level of one gene. Some dimensions contribute more to the variation between patients than other dimensions. The first principal component is the axis that captures most variation between patients. A number of genes have a projection on this axis, and the principal component method can tell you how much each gene contributes to the axis. The genes that contribute the most show the most variation between patients. Can this method be used for selecting diagnostic genes? Yes, but it is not necessarily the best method, because variation, as we have seen in Section 3.6, can be due to noise as well as true difference in expression. Besides, how do we know how many genes to pick? The t-test and ANOVA, mentioned in Section 3.6, are more suited to this task. They will, based on a replication of measurement in each patient class, tell you which genes vary between patient classes and give you the probability of false positives at the cutoff you choose.

So a more useful application of principal component analysis would be to visualize genes that have been found by a t-test or ANOVA to be significantly regulated. This visualization may give you ideas for further analysis of the data.

4.2 SUPERVISED PRINCIPAL COMPONENT ANALYSIS

An alternative to performing PCA on genes that have been selected to be relevant to your problem is to use supervised PCA. Partial least squares (PLS) is a method of supervised PCA for microarray analysis (Nguyen and Rocke, 2002a–c). PLS uses the labeled data as input and identifies the variance associated with the class distinction. This approach has worked well for classification purposes.

4.3 INDEPENDENT COMPONENT ANALYSIS

Independent component analysis (ICA) is related to principal component analysis. The main difference between ICA and PCA is that ICA extracts statistically independent components that are non-Gaussian, whereas components extracted by PCA can be pure noise (Gaussian). One advantage of ICA in this context is that it is possible to skip the gene selection step that is often used before PCA to reduce noise or uninteresting variation. Instead, we can just eliminate genes that have no variation at all, either because they are not expressed in any of the samples, or because their expression is constant. Independent component analysis for gene expression data has compared favorably to both PCA and PLS (Carpentier et al., 2004).

The independent components fulfill the following relationship:

$$X = SA,$$

where X is the transposed expression matrix of genes versus samples, S contains the independent components, and A is a linear mixing matrix.

Independent component analysis is computationally much more complex than PCA: it uses iteration from a random starting point to arrive at the final components that fulfill the relationship. Thus it is not guaranteed to find the same solution every time it is run.

4.4 EXAMPLE I: PCA ON SMALL DATA MATRIX

Let us look at a simple example to visualize the problem. We have the data matrix shown in Table 4.1.

TABLE 4.1 Expression Readings of Four Genes in Six Patients

Gene	Patient					
	N_1	N_2	A_1	A_2	B_1	B_2
a	90	110	190	210	290	310
b	190	210	390	410	590	610
c	90	110	110	90	120	80
d	200	100	400	90	600	200

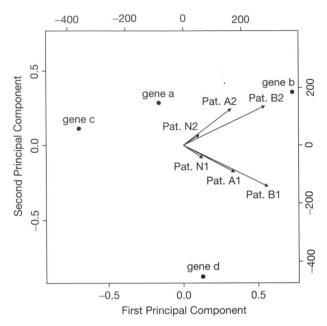

Figure 4.2 *Principal component analysis of data shown in Table 4.1. Note that patients and genes use different coordinate systems in this plot, hence the different scales on the axes.*

It consists of four genes measured in six patients. If we perform a principal component analysis on these data (the details of the computation are shown in Section 10.5), we get the biplot shown in Figure 4.2. A biplot is a plot designed to visualize both points and axes simultaneously. Here we have plotted the four genes as points in two dimensions, the first two principal components. It can be seen that genes a, c, and b differ a lot in the first dimension (they vary from about -400 to $+400$), while they differ little in the second dimension. Gene d, however, is separated from the other genes in the second dimension (it has a value of about -400 in the second dimension).

Indicated as arrows are the projections of the six patient axes on the two first principal components. Start with Patient B_1. This patient has a large projection (about 0.5) on the first principal component, and a smaller projection on the second principal component (about -0.3). The lengths of the patient vectors indicate how much they contribute to each axis and their directions indicate in which way they contribute. The first principal component consists mainly of Patient category B, where expression differs most. Going back to the genes, it can be seen that they are ranked according to average expression level in the B patients along this first principal component: genes c, a, d, and b.

The second principal component divides genes into those that are higher in Patient B_2 than in Patient B_1 (genes c, a, and b), and gene d, which is lower in Patient B_2 than in Patient B_1. On the vector projections of the patient axes on this component it can be seen that they have been divided into those with subcategory 1 (Patients N_1, A_1, B_1), which all have a positive projection, and those with subcategory 2 (Patients

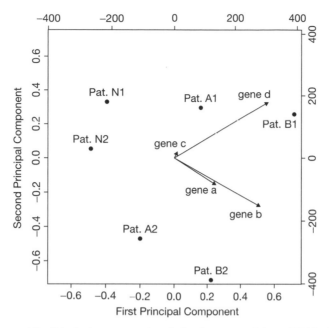

Figure 4.3 *Principal component analysis of transposed data of Table 4.1.*

N_2, A_2, B_2), which all have a negative projection. So the second principal component simply compares expression in subcategories.

We can also do the principal component analysis on the reverse (*transposed*) matrix (transposition means to swap rows with columns). Figure 4.3 shows a biplot of patients along principal components that consist of those genes that vary most between patients. First, it can be seen that there has been some grouping of patients into categories. Categories can be separated by two parallel lines. By looking at the projection of the gene vectors we can see that gene *b* and gene *d*, those that vary most, contribute most to the two axes. Now, if we wanted to use this principal component analysis to select genes that are diagnostic for the three categories, we might be tempted to select gene *b* and gene *d* because they contribute most to the first principal component. This would be a mistake, however, because gene *d* just shows high variance that is not correlated to category at all. The ANOVA, described in Section 3.6, would have told us that gene *b* and gene *a* are the right genes to pick as diagnostic genes for the disease.

4.5 EXAMPLE II: PCA ON REAL DATA

Figure 4.4 shows results of a PCA on real data. The R package was used on normalized HIV data as described in Section 10.5. It can be seen that a diagonal line separates the three control samples from the three HIV-infected samples. Thus HIV constitutes one of the major sources of variation in this experiment, and this variation is captured in the first two principal components. In fact, the first principal component perfectly separates control samples from HIV samples.

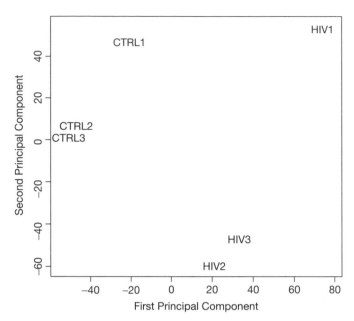

Figure 4.4 *Principal component analysis on real data from in vitro HIV infection experiment. Six samples are plotted along the first two principal components.*

4.6 SUMMARY

Principal component analysis is a way to reduce your multidimensional data to a single $x-y$ graph. You may be able to spot important trends in your data from this one visualization alone.

FURTHER READING

Singular Value Decomposition

Alter, O., Brown, P. O., and Botstein, D. (2000). Singular value decomposition for genome-wide expression data processing and modeling. *Proc. Natl. Acad. Sci. USA* 97:10101–10106.

Holter, N. S., Mitra, M., Maritan, A., Cieplak, M., Banavar, J. R., and Fedoroff, N.V. (2000). Fundamental patterns underlying gene expression profiles: simplicity from complexity. *Proc. Natl. Acad. Sci. USA* 97:8409–8414.

Wall, M. E., Dyck, P. A., and Brettin, T. S. (2001). SVDMAN—singular value decomposition analysis of microarray data. *Bioinformatics* 17:566–568.

Principal Component Analysis

Dysvik, B., and Jonassen, I. (2001). J-Express: exploring gene expression data using Java. *Bioinformatics* 17:369–370. (Software available at http://www.ii.uib.no/~bjarted/jexpress/.)

Raychaudhuri, S., Stuart, J. M., and Altman, R. B. (2000). Principal components analysis to summarize microarray experiments: application to sporulation time series. *Pacific Symposium on Biocomputing* 2000:455–466. (Available online at http://psb.stanford.edu.)

Xia, X., and Xie, Z. (2001). AMADA: analysis of microarray data. *Bioinformatics* 17:569–570.

Xiong, M., Jin, L., Li, W., and Boerwinkle, E. (2000). Computational methods for gene expression-based tumor classification. *Biotechniques* 29:1264–1268.

Partial Least Squares

Ghosh, D. (2003). Penalized discriminant methods for the classification of tumors from gene expression data. *Biometrics* 59(4):992–1000.

Perez-Enciso, M., and Tenenhaus, M. (2003). Prediction of clinical outcome with microarray data: a partial least squares discriminant analysis (PLS-DA) approach. *Hum. Genet.* 112 (5-6):581–592.

Tan, Y., Shi, L., Tong, W., Gene, H., and Wang, C. (2004). Multi-class tumor classification by discriminant partial least squares using microarray gene expression data and assessment of classification models. *Comput. Biol. Chem.* 28(3):235–244.

Independent Component Analysis

Chiappetta, P., Roubaud, M. C., and Torresani, B. (2004). Blind source separation and the analysis of microarray data. *J. Comput. Biol.* 11(6):1090–1109.

Lee, S. I., and Batzoglou, S. (2003). Application of independent component analysis to microarrays. *Genome Biol.* 4(11):R76. Epub 2003 Oct 24.

Liebermeister, W. (2002). Linear modes of gene expression determined by independent component analysis. *Bioinformatics* 18(1):51–60.

Martoglio, A. M., Miskin, J. W., Smith, S. K., and MacKay, D. J. (2002). A decomposition model to track gene expression signatures: preview on observer-independent classification of ovarian cancer. *Bioinformatics* 18(12):1617–1624.

Saidi, S. A., Holland, C. M., Kreil, D. P., MacKay, D. J., Charnock-Jones, D. S., Print, C. G., and Smith, S. K. (2004). Independent component analysis of microarray data in the study of endometrial cancer. *Oncogene* 23(39):6677–6683.

Correspondence Analysis

Fellenberg, K., Hauser, N. C., Brors, B., Neutzner, A., Hoheisel, J. D., and Vingron, M. (2001). Correspondence analysis applied to microarray data. *Proc. Natl. Acad. Sci. USA* 98:10781–10786.

Gene Shaving Uses PCA to Select Genes with Maximum Variance

Hastie, T., Tibshirani, R., Eisen, M. B., Alizadeh, A., Levy, R., Staudt, L., Chan, W. C., Botstein, D., and Brown, P. (2000). Gene shaving as a method for identifying distinct sets of genes with similar expression patterns. *Genome Biol.* 1:1–21.

5

Cluster Analysis

If you have just one experiment and a control, your first data analysis will limit itself to a list of regulated genes ranked by the magnitude of up- and downregulation, or ranked by the significance of regulation determined in a t-test.

Once you have more experiments—measuring the same genes under different conditions, in different mutants, in different patients, or at different time points during an experiment—it makes sense to group the significantly changed genes into clusters that behave similarly over the different conditions. It is also possible to use clustering to group patients into those who have a similar disease. Clustering is often used to discover new subtypes of cancer in this way.

5.1 HIERARCHICAL CLUSTERING

Think of each gene as a vector of N numbers, where N is the number of experiments or patients. Then you can plot each gene as a point in N-dimensional hyperspace. You can then calculate the distance between two genes as the Euclidean distance between their respective data points (as the square root of the sum of the squared distances in each dimension).

This can be visualized using a modified version of the small example dataset applied in previous chapters (Table 5.1). The measured expression level of the five genes can be plotted in just two of the patients using a standard $x-y$ coordinate system (Figure 5.1 left).

You can calculate the distance between all genes (producing a *distance matrix*), and then it makes sense to group those genes together that are closest to each other in space. The two genes that are closest to each other, b and d, form the first cluster (Figure 5.1 left). Genes a and c are separated by a larger distance, and they form a

Cancer Diagnostics with DNA Microarrays, By Steen Knudsen
Copyright © 2006 John Wiley & Sons, Inc.

TABLE 5.1   Expression Readings of Five Genes in Two Patients

	Patient	
Gene	N_1	A_1
a	90	190
b	190	390
c	90	110
d	200	400
e	150	200

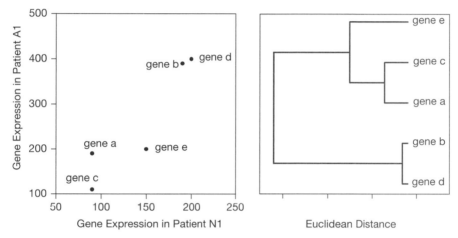

Figure 5.1 *Hierarchical clustering of genes based on their Euclidean distance visualized by a rooted tree. Note that it is possible to reorder the leaves of the tree by flipping the branches at any node without changing the information in the tree.*

cluster as well (Figure 5.1 left). If the separation between a gene and a cluster comes within the distance as you increase it, you add that gene to the cluster. Gene *e* is added to the cluster formed by *a* and *c*. How do you calculate the distance between a point (gene) and a cluster? You can calculate the distance to the *nearest neighbor* in the cluster (gene *a*). This is called single linkage. You can also calculate the distance to the farthest member of the cluster (complete linkage), but it is best to calculate the distance to the point that is in the middle of the existing members of the cluster (centroid, similar to UPGMA or average linkage method).

When you have increased the distance to a level where all genes fall within that distance, you are finished with the clustering and can connect the final clusters. You have now performed a *hierarchical agglomerative clustering*. There are computer algorithms available for doing this (see Section 10.4).

A real example is shown in Figure 5.2, where bladder cancer patients were clustered based on Affymetrix GeneChip expression measurements from a bladder biopsy. It is seen in the figure that the clustering groups superficial tumors together and groups invasive tumors together.

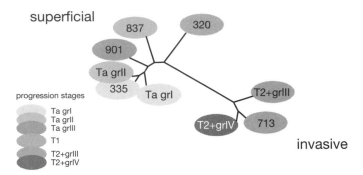

Figure 5.2   *Hierarchical clustering of bladder cancer patients using an unrooted tree. The clustering was based on expression measurements from a DNA array hybridized with mRNA extracted from a biopsy. Numbers refer to patients and the severity of the disease is indicated by a grayscale code (see color insert). (Christopher Workman, based on data published in Thykjaer et al., 2001.)*

Hierarchical clustering only fails when you have a large number of genes (several thousand). Calculating the distances between all of them becomes time consuming. Removing genes that show no significant change in any experiment is one way to reduce the problem. Another way is to use a faster algorithm, like K-means clustering.

5.2   *K*-MEANS CLUSTERING

In K-means clustering, you skip the calculation of distances between all genes. You decide on the number of clusters you want to divide the genes into, and the computer then randomly assigns each gene to one of the K clusters. Now it will be comparatively fast to calculate the distance between each gene and the center of each cluster (*centroid*). If a gene is actually closer to the center of another cluster than the one it is currently assigned to, it is reassigned to the closer cluster. After assigning all genes to the closest cluster, the centroids are recalculated. After a number of iterations of this, the cluster centroids will no longer change, and the algorithm stops. This is a very fast algorithm, but it will give you only the number of clusters you asked for and not show their relation to each other as a full hierarchical clustering will do. In practice, K-means is useful if you try different values of K.

If you try the K-means clustering on the expression data used for hierarchical clustering shown in Figure 5.1, with $K = 2$, the algorithm may find the solution in just one iteration (Figure 5.3).

5.3   SELF-ORGANIZING MAPS

There are other methods for clustering, but hierarchical and K-means cover most needs. One method that is available in a number of clustering software packages is self-organizing maps (SOM) (Kohonen, 1995). SOM is similar to K-means, but clusters are ordered on a low-dimensional structure, such as a grid. The advantage over K-means is that neighboring clusters in this grid are more related than clusters that are not neighbors. So it results in an ordering of clusters that is not performed in K-means.

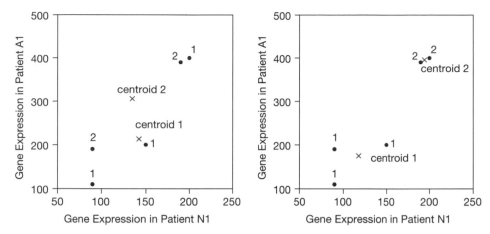

Figure 5.3 *K-means clustering of genes based on their Euclidean distance. First, genes are randomly assigned to one of the two clusters in K: 1 or 2 (left). The centroids of each cluster are calculated. Genes are then reassigned to another cluster if they are closer to the centroid of that cluster (right). In this simple example, the final solution is obtained after just one iteration (right).*

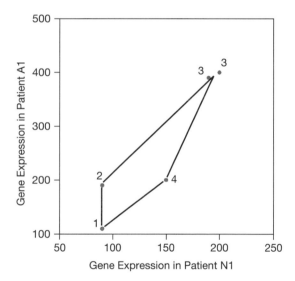

Figure 5.4 *SOM clustering of genes into a two-by-two grid, resulting in four clusters.*

Figure 5.4 shows how a SOM clustering could fit a two-by-two grid on the data of our example (the cluster centers are at each corner of the grid). That would result in four clusters, three of them with one member only and one cluster with two members.

5.4 DISTANCE MEASURES

In addition to calculating the Euclidean distance, there are a number of other ways to calculate distance between two genes. When these are combined with different ways of normalizing your data, the choice of normalization and distance measure can become rather confusing. Here I will attempt to show how the different distance measures relate to each other and what effect, if any, normalization of the data has. Finally, I will suggest a good choice of distance measure for expression data.

The Euclidean distance between two points a and b in N-dimensional space is defined as

$$\sqrt{\sum_{i=1}^{N}(a_i - b_i)^2},$$

where i is the index that loops over the dimensions of N, and the $\sum$ sign indicates that the squared distances in each dimension should be summed before taking the square root of those sums. Figure 5.5 shows the Euclidean distance between two points in two-dimensional space.

Instead of calculating the Euclidean distance, you can also calculate the angle between the vectors that are formed between the data point of the gene and the center of the coordinate system. For gene expression, *vector angle* (Figure 5.5) often performs better because the trend of a regulation response is more important than its magnitude. Vector angle α between points a and b in N-dimensional space is calculated as

$$\cos \alpha = \frac{\sum_{i=1}^{N} a_i b_i}{\sqrt{\sum_{i=1}^{N} a_i^2} \sqrt{\sum_{i=1}^{N} b_i^2}}.$$

Finally, a widely used distance metric is the Pearson correlation coefficient:

$$\text{PearsonCC} = \frac{\sum_{i=1}^{N} (a_i - \overline{a})(b_i - \overline{b})}{\sqrt{\sum_{i=1}^{N} (a_i - \overline{a})^2} \sqrt{\sum_{i=1}^{N} (b_i - \overline{b})^2}}.$$

You can see that the only difference between vector angle and Pearson correlation is that the means ($\overline{a}$ and $\overline{b}$) have been subtracted before calculating the Pearson correlation. So taking the vector angle of a means-normalized dataset (each gene has been

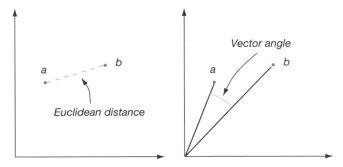

Figure 5.5 *Euclidean distance and vector angle between points a and b in two-dimensional space.*

centered around its mean expression value over all conditions) is the same as taking the Pearson correlation.

An example will illustrate this point. Let us consider two genes, a and b, that have the expression levels $a = (1, 2, 3, 4)$ and $b = (2, 4, 6, 8)$ in four experiments. They both show an increasing expression over the four experiments, but the magnitude of response differs. The Euclidean distance is 5.48, while the vector angle distance $(1 - \cos \alpha)$ is zero and the Pearson distance (1 − Pearson CC) is zero. I would say that because the two genes show exactly the same trend in the four experiments, the vector angle and Pearson distance make more sense in a biological context than the Euclidean distance.

5.4.1   Example: Comparison of Distance Measures

Let us try the different distance measures on our little example of four genes from six patients (Table 5.2). We can calculate the Euclidean distances (see Section 10.4 for details on how to do this) between the four genes. The pairwise distances between all genes can be shown in a *distance matrix* (Table 5.3) where the distance between gene a and a is zero, so the pairwise identities form a diagonal of zeros through the matrix. The triangle above the diagonal is a mirror image of the triangle below the diagonal because the distance between genes a and b is the same as the distance between genes b and a.

This distance matrix is best visualized by clustering as shown in Figure 5.6, where it is compared with clustering based on vector angle distance (Table 5.4) and a tree based on Pearson correlation distances (Table 5.5).

The clustering in Figure 5.6 is nothing but a two-dimensional visualization of the four-by-four distance matrix. What does it tell us? It tells us that vector angle distance is the best way to represent gene expression responses. Genes a, b, and d all have increasing expression over the three patient categories, only the magnitude of the response and the error between replicates differ. The vector angle clustering has

TABLE 5.2   Expression Readings of Four Genes in Six Patients

Gene	Patient					
	N_1	N_2	A_1	A_2	B_1	B_2
a	90	110	190	210	290	310
b	190	210	390	410	590	610
c	90	110	110	90	120	80
d	200	100	400	90	600	200

TABLE 5.3   Euclidean Distance Matrix Between Four Genes

Gene	Gene			
	a	b	c	d
a	0.00	5.29	3.20	4.23
b	5.29	0.00	8.38	5.32
c	3.20	8.38	0.00	5.84
d	4.23	5.32	5.84	0.00

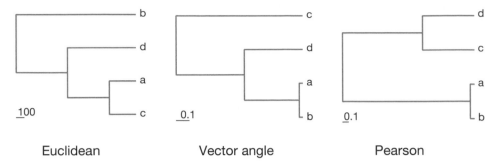

Figure 5.6 *Hierarchical clustering of distances (with three different distance measures) between genes in the example.*

TABLE 5.4   Vector Angle Distance Matrix
Between Four Genes

| | Gene | | | |
Gene	*a*	*b*	*c*	*d*
a	0.00	0.02	0.42	0.52
b	0.02	0.00	0.41	0.50
c	0.42	0.41	0.00	0.51
d	0.52	0.50	0.51	0.00

TABLE 5.5   Pearson Distance Matrix Between
Four Genes

| | Gene | | | |
Gene	*a*	*b*	*c*	*d*
a	0.00	0.06	1.45	1.03
b	0.06	0.00	1.43	0.98
c	1.45	1.43	0.00	0.83
d	1.03	0.98	0.83	0.00

captured this trend perfectly, grouping *a* and *b* close together and *d* nearby. Euclidean distance has completely missed this picture, focusing only on absolute expression values, and placed genes *a* and *b* furthest apart. Pearson correlation distance has done a pretty good job, capturing the close biological proximity of genes *a* and *b*, but it has normalized the data too heavily and placed gene *d* closest to gene *c*, which shows no trend in the disease at all.

It is also possible to cluster in the other dimension, clustering patients instead of genes. Instead of looking for genes that show a similar transcriptional response to the progression of a disease, we are looking for patients who have the same *profile* of expressed genes. If two patients have exactly the same stage of a disease we hope that this will be reflected in identical expression of a number of key genes. Thus we are not interested in the genes that are not expressed in any patient, are unchanged between patients, or show a high error. So it makes sense to remove those genes before clustering patients. But that requires applying a *t*-test or ANOVA on the data and in

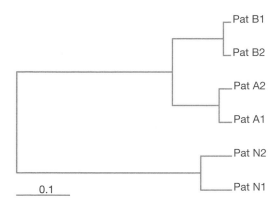

Pat B1
Pat B2
Pat A2
Pat A1
Pat N2
0.1
Pat N1

Figure 5.7 *Hierarchical clustering of vector angle distances between patients in the example.*

order to do so you have to put the data into the categories you presume the patients fall into. That could obscure trends in the data that you have not yet considered. In practice, you could try clustering both on the full data and on data cleaned by t-test, ANOVA, or another method.

The data in our little example contains too much noise from gene c to cluster on the complete set of genes. If an ANOVA is run on the three patient categories (Section 3.6) that only leaves genes a and b, and the hierarchical vector angle distance clustering based on those two genes can be seen in Figure 5.7. But remember that these two genes were selected explicitly for the purpose of separating the three patient categories.

5.5 TIME-SERIES ANALYSIS

When your data consist of samples from different time points in an experiment, this presents a unique situation for analysis. There are two fundamentally different ways of approaching the analysis. You can take replicate samples from each time point and use statistical methods such as t-test or ANOVA to identify genes that are expressed differentially over the time series. This approach does not rely on any assumptions about the spacing between time points, or the behavior of gene expression over time.

Another way of approaching the analysis is to assume that there is some relationship between the time points. For example, you can assume that there is a linear relationship between the samples, so that genes increase or decrease in expression in a linear manner over time. In that case you can use linear modeling as a statistical analysis tool.

Another possible relationship is a sine wave for cyclical phenomena. The sine wave has been used to analyze cell cycle experiments in yeast (Spellman et al., 1998).

If you have no prior expectation on the response in your data, clustering may be the most powerful way of discovering temporal relationships.

5.6 GENE NORMALIZATION

The difference between vector angle distance and Pearson correlation comes down to a means normalization. There are two other common ways of normalizing the expression level of a gene—length normalization and SD normalization:

- Means: Calculate mean and subtract from all numbers.
- Length: Calculate length of gene vector and divide all numbers by that length.
- SD: Calculate standard deviation and divide all numbers by it.

For each of these normalizations it is important to realize that it is performed on each gene in isolation; the information from other genes is not taken into account. Before you perform any of these normalizations, it is important that you answer the following question. Why do you want to normalize the data in that way? Remember, you have already normalized the chips, so expression readings should be comparable. In general, normalization affects Euclidean distances to a large extent, it affects vector angles to a much smaller extent, and it hardly ever affects Pearson distances because the latter metric is normalized already. My suggestion for biological data is to use vector angle distance on nonnormalized expression data for gene clustering.

5.7   VISUALIZATION OF CLUSTERS

Clusters are traditionally visualized with trees (Figures 5.1, 5.2, 5.6, and 5.7). Note that information is lost in going from a full distance matrix to a tree visualization of it. Different trees can represent the same distance matrix.

In DNA chip analysis it has also become common to visualize the gene vectors by representing the expression level or fold change in each experiment with a color-coded matrix (Section 10.4).

5.8   SUMMARY

Cluster analysis groups genes according to how they behave in experiments. For gene expression, measuring similarity of gene expression using the vector angle between expression profiles of two genes makes the most sense.

Normalization of your data matrix (of genes versus experiments) can be performed in either of two dimensions. If you normalize columns you normalize the total expression level of each experiment. A normalization of experiments to yield the same sum of all genes is referred to in this book as scaling and is described in Section 3.1. Such a normalization is essential before comparison of experiments, but a multifactor scaling with a spline or a polynomial is even better.

Normalization of genes in the other dimension may distort the scaling of experiments that you have performed (if you sum the expression of all genes in an experiment after a gene normalization, it will no longer add up to the same number). Also, normalization of genes before calculating vector angle is usually not necessary. Therefore, Pearson correlation is not quite as good a measure of similarity as vector angle.

FURTHER READING

Clustering Methods and Cluster Reliability

Ben-Hur, A., Elisseeff, A., and Guyon, I. (2002). A stability based method for discovering structure in clustered data. *Pacific Symposium on Biocomputing* **2002**:6–17. (Available online at http://psb.stanford.edu.)

De Smet, F., Mathys, J., Marchal, K., Thijs, G., De Moor, B., and Moreau, Y. (2002). Adaptive quality-based clustering of gene expression profiles. *Bioinformatics* 18(5):735–746.

Getz, G., Levine, E., and Domany, E. (2000). Coupled two-way clustering analysis of gene microarray data. *Proc. Natl. Acad. Sci. USA* 97:12079–12084.

Hastie, T., Tibshirani, R., Eisen, M. B., Alizadeh, A., Levy, R., Staudt, L., Chan, W. C., Botstein, D., and Brown, P. (2000). Gene shaving as a method for identifying distinct sets of genes with similar expression patterns. *Genome Biol.* 1:1–21.

Herrero, J., Valencia, A., and Dopazo, J. (2001). A hierarchical unsupervised growing neural network for clustering gene expression patterns. *Bioinformatics* 17:126–136.

Kerr, M. K., and Churchill, G. A. (2001). Bootstrapping cluster analysis: assessing the reliability of conclusions from microarray experiments. *Proc. Natl. Acad. Sci. USA* 98:8961–8965.

Michaels, G. S., Carr, D. B., Askenazi, M., Fuhrman, S., Wen, X., and Somogyi, R. (1998). Cluster analysis and data visualization of large-scale gene expression data. *Pacific Symposium on Biocomputing* 3:42–53. (Available online at http://psb.stanford.edu.)

Sasik, R., Hwa, T., Iranfar, N., and Loomis, W. F. (2001). Percolation clustering: a novel algorithm applied to the clustering of gene expression patterns in dictyostelium development. *Pacific Symposium on Biocomputing* 6:335–347. (Available online at http://psb.stanford.edu.)

Tamayo, P., Slonim, D., Mesirov, J., Zhu, Q., Kitareewan, S., Dmitrovsky, E., Lander, E. S., and Golub, T. R. (1999). Interpreting patterns of gene expression with self-organizing maps: methods and application to hematopoietic differentiation. *Proc. Natl. Acad. Sci. USA* 96:2907–2912.

Tibshirani, R., Walther, G., Botstein, D., and Brown, P. (2000). Cluster validation by prediction strength. Technical report. Statistics Department, Stanford University. (Manuscript available at http://www-stat.stanford.edu/~tibs/research.html.)

Xing, E. P., and Karp, R. M. (2001). CLIFF: clustering of high-dimensional microarray data via iterative feature filtering using normalized cuts. *Bioinformatics* 17(Suppl 1):S306–S315.

Yeung, K. Y., Fraley, C., Murua, A., Raftery, A. E., and Ruzzo, W. L. (2001a). Model-based clustering and data transformations for gene expression data. *Bioinformatics* 17:977–987.

Yeung, K. Y., Haynor, D. R., and Ruzzo, W. L. (2001b). Validating clustering for gene expression data. *Bioinformatics* 17:309–318.

6

Molecular Classifiers for Cancer

Perhaps the most promising application of DNA microarrays for expression profiling is toward classification: in particular, in medicine, where DNA microarrays may define profiles that characterize specific phenotypes (diagnosis), predict a patient's clinical outcome (prognosis), or predict which treatment is most likely to benefit the patient (tailored treatment).

The only limitation seems to be the fact that a sample of the diseased tissue is required for the chip. That limits the application to diseases that affect cells that can easily be obtained: blood disease where a blood sample can easily be obtained, or tumors where a biopsy is routinely obtained or the entire tumor is removed during surgery. Consequently, DNA microarrays have in the past few years been applied to almost any cancer type known to man, and in most cases it has been possible to distinguish clinical phenotypes based on the array alone. Where data on long-term outcome has been available, it has also been possible to predict that outcome to a certain extent using DNA arrays.

The key to the success of DNA microarrays in this field is that it is not necessary to understand the underlying molecular biology of the disease. Rather, it is a purely statistical exercise in linking a certain pattern of expression to a certain diagnosis or prognosis. This is called classification and it is a well established field in statistics, from where we can draw upon a wealth of methods suitable for the purpose. This chapter briefly explains some of the more common methods that have been applied to DNA microarray classification with good results.

6.1 SUPERVISED VERSUS UNSUPERVISED ANALYSIS

The hundreds of papers that have been published showing application of DNA micro-arrays to the diagnosis or prognosis of disease will be reviewed in later chapters that

Cancer Diagnostics with DNA Microarrays, By Steen Knudsen
Copyright © 2006 John Wiley & Sons, Inc.

are specific to the disease. They typically make use of one of two fundamentally different approaches: an *unsupervised* analysis of the data or a *supervised* analysis of the data. In the unsupervised analysis patients are clustered without regard to their class label (their clinical diagnosis). After the clustering has been performed, the clusters are compared to clinical data to look for correlations. Not only can this approach be used to identify subgroups of patients with a certain clinical diagnosis or prognosis, it can also be used to discover new subgroups hitherto unrecognized in clinical practice. As will be described in the chapters on specific diseases, the unsupervised approach has been used with success to discover new subtypes of well-described diseases.

The supervised approach, on the other hand, uses the class label (clinical diagnosis) to group the patients and then uses statistical methods to look for differences between the groups that can be used to distinguish them. Looking for distinguishing features is called feature selection. Once features have been selected a classifier can be built that can classify a new sample (patient) based on the features. This is the key approach to molecular classification. Compared to the unsupervised analysis it is typically a more efficient search for the best classifier, but there are two important limitations. First, supervised analysis cannot be used to discover new subgroups. Second, if there are errors in the class labels (clinical diagnosis), it will affect classifier accuracy if it is not discovered. For that reason it is crucial to perform both a supervised and an unsupervised analysis of your data.

6.2   FEATURE SELECTION

Feature selection is a very critical issue where it is easy to make mistakes. You can build your classifier using the expression of all genes on a chip as input or you can select the features (genes) that seem important for the classification problem at hand. As described below, it is easy to overfit the selected genes to the samples on which you are basing the selection. Therefore it is *crucial* to validate the classifier on a set of samples that were not used for feature selection.

There are many ways in which to select genes for your classifier: the simplest is to use the *t*-test or ANOVA described in this book to select genes differing significantly in expression between the different diagnosis categories or prognosis categories you wish to predict. In comparison experiments with the many other feature selection methods for DNA microarrays that have been published, these simple approaches have often performed well.

The *t*-test and similar approaches have been widely used for feature selection, but there is one common mistake that is often seen in published studies. Often both the training set and the test set used for validation are used for feature selection. In that case the performance of the classifier will be overestimated because the features of the test set have been used to build the classifier.

It is also possible to select features based on whether they improve classification performance or not. These methods (referred to as embedded or wrapper methods) have the advantage of eliminating redundant features and taking into account interactions among features.

6.3   VALIDATION

If you have two cancer subtypes and you run one chip on each of them, can you then use the chip to classify the cancers into the two subtypes? With 6000 genes or more, easily. You can pick any gene that is expressed in one subtype and not in the other and use that to classify the two subtypes.

What if you have several cancer tissue specimens from one subtype and several specimens from the other subtype? The problem becomes only slightly more difficult. You now need to look for genes that all the specimens from one subtype have in common and are absent in all the specimens from the other subtype.

The problem with this method is that you have just selected genes to fit your data—you have not extracted a *general* method that will *classify any specimen of one of the subtypes that you are presented with after building your classifier.*

In order to build a *general* method, you have to observe several basic rules (Simon et al., 2003):

- Avoid overfitting data. Use fewer estimated parameters than the number of specimens on which you are building your model. Every time you use a test set for optimizing some parameter (like the optimal number of features to use) you have to reserve an independent test set for validation. Independent means that the test set was not used to estimate any parameter of the classifier.
- Validate your method by testing it on an independent dataset that was not used for building the model. (If your dataset is very small, you can use *cross-validation*, where you subdivide your dataset into test and training several times. If you have ten examples, there are ten ways in which to split the data into a training set of nine and a test set of one. That is called a tenfold cross-validation. That is also called a leave-one-out cross-validation or LOOCV.)

6.4   CLASSIFICATION SCHEMES

Most classification methods take as input points in space where each point corresponds to one patient or sample and each dimension in space corresponds to the expression of a single gene. The goal is then to classify a sample based on its position in space relative to the other samples and their known classes. As such, this method is related to the principal component analysis and clustering described elsewhere in this book. A key difference is that those methods are unsupervised; they do not use the information of the class relationship of each sample. Classification is supervised; the class relationship of each sample is used to build the classifier.

6.4.1   Nearest Neighbor

The simplest form of classifier is called a nearest neighbor classifier (Section 10.5.1.6) (Fix and Hodges, 1951; Dudoit et al., 2000a). The general form uses k nearest neighbors (KNNs) and proceeds as follows: (1) plot each patient in space according to the expression of the genes; (2) for each patient, find the k nearest neighbors according to the distance metric you choose; (3) predict the class by majority vote, that is, the class that is most common among the k neighbors. If you use only odd values of k, you avoid the situation of a vote tie. Otherwise, vote ties can be broken by a random generator. The value of k can be chosen by cross-validation to minimize the prediction error on a labeled test set. (See Figure 6.1.)

If the classes are well separated in an initial principal component analysis (Section 4.4) or clustering, nearest neighbor classification will work well. If the classes

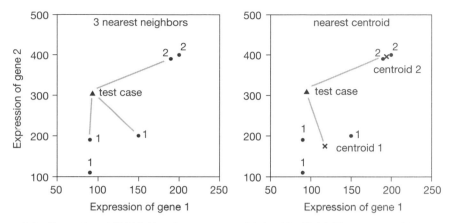

Figure 6.1   *Illustration of KNN (left) and nearest centroid classifier (right). The majority of the 3 nearest neighbors (left) belong to class 1; therefore we classify the test case as belonging to class 1. The nearest centroid (right) is that of class 1; therefore we classify the test case as belonging to class 1.*

are not separable by principal component analysis, it may be necessary to use more advanced classification methods, such as neural networks or support vector machines. The k nearest neighbor classifier will work both with and without feature selection. Either you can use all genes on the chip or you can select informative genes with feature selection.

6.4.2   Nearest Centroid

Related to the nearest neighbor classifier is the nearest centroid classifier. Instead of looking at only the nearest neighbors it uses the centroids (center points) of all members of a certain class. The patient to be classified is assigned the class of the nearest centroid.

6.4.3   Neural Networks

If the number of examples is sufficiently high (between 50 and 100), it is possible to use a more advanced form of classification. Neural networks (Section 10.5.1.7) simulate some of the logic that lies beneath the way in which brain neurons communicate with each other to process information. Neural networks *learn* by adjusting the strengths of connections between them. In computer-simulated artificial neural networks, an algorithm is available for learning based on a learning set that is presented to the software. The neural network consists of an input layer where examples are presented, and an output layer where the answer, or classification category, is output. There can be one or more hidden layers between the input and output layers.

To keep the number of adjustable parameters in the neural network as small as possible, it is necessary to reduce the dimensionality of array data before presenting it to the network. Khan et al. (2001) used principal component analysis and presented only the most important principal components to the neural network input layer. They then used an ensemble of cross-validated neural networks to predict the cancer class of patients.

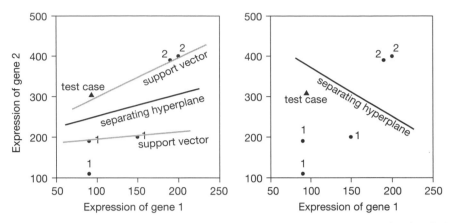

Figure 6.2 *Illustration of support vector machine (left) and neural network classifier (right). Both define a separating hyperplane that can be defined in higher dimensions and can be more complex than what is shown here.*

6.4.4 Support Vector Machine

Another type of classifier is the support vector machine (Brown et al., 2000; Dudoit et al., 2000a), a machine learning approach (Figure 6.2). It is well suited to the dimensionality of array data. R code for implementing support vector machines can be found in the e1071 package at the R project web site (www.r-project.org).

6.5 PERFORMANCE EVALUATION

There are a number of different measures for evaluating the performance of your classifier on an independent test set. First, if you have a binary classifier that results in only two classes (e.g., cancer or normal), you can use Matthews' correlation coefficient (Matthews, 1975) to measure its performance:

$$CC = \frac{(TP \times TN) - (FP \times FN)}{\sqrt{(TP + FN)(TP + FP)(TN + FP)(TN + FN)}},$$

where TP is the number of true positive predictions, FP is the number of false positive predictions, TN is the number of true negative predictions, and FN is the number of false negative predictions. A correlation coefficient of 1 means perfect prediction, whereas a correlation coefficient of zero means no correlation at all (that could be obtained from a random prediction).

When the output of your classifier is continuous, such as that from a neural network, the numbers TP, FP, TN, and FN depend on the threshold applied to the classification. In that case you can map out the correlation coefficient as a function of the threshold in order to select the threshold that gives the highest correlation coefficient. A more common way to show how the threshold affects performance, however, is to produce a ROC curve (receiver operating characteristics). In a ROC curve you plot the sensitivity (TP/(TP + FN)) versus the false positive rate (FP/(FP + TN)). One way of comparing the performance of two different classifiers is then to compare the area under the ROC curve. The larger the area, the better the classifier.

6.6   EXAMPLE I: CLASSIFICATION OF SRBCT CANCER SUBTYPES

Khan et al. (2001) have classified small, round blue cell tumors (SRBCT) into four classes using expression profiling and kindly made their data available on the World Wide Web (http://www.thep.lu.se/pub/Preprints/01/lu_tp_01_06_supp.html). We can test some of the classifiers mentioned in this chapter on their data.

First, we can try a *k* nearest neighbor classifier (see Section 10.5.1.6 for details). Using the full dataset of 2000 genes, and defining the nearest neighbors in the space of 63 tumors by Euclidean distance, a $k = 3$ nearest neighbor classifier classifies 61 of the 63 tumors correctly in a leave-one-out cross-validation (each of the 63 tumors is classified in turn, using the remaining 62 tumors as a reference set).

We can also train a neural network (see Section 10.5.1.7 for details) to classify tumors into four classes based on principal components. Twenty feed-forward neural networks are trained on 62 tumors and used to predict the class of the 63rd tumor based on a committee vote among the twenty networks (this is a leave-one-out cross-validation). Figure 6.3 shows the first two principal components from a principal component analysis. The first ten principal components from a principal component analysis are used as the input for the neural network for each tumor, and four neurons are used for output, one for each category. Interestingly, two of the classes are best predicted with no hidden neurons, and the other two classes are best predicted with a hidden layer of two neurons. Using this setup, the neural networks classify 62 of the 63 tumors correctly. But of course, these numbers have to be validated on an independent, blind set as done by Khan et al. (2001).

6.7   A NETWORK APPROACH TO MOLECULAR CLASSIFICATION

The approach to classifying biological samples with DNA microarrays described so far is based on the one gene–one function paradigm. Thus in order to build a classifier that can classify biological samples into two or more categories based on their gene expression, a typical step is to select a list of genes, the activity of which can be used as input to the classification algorithm (Figure 6.4).

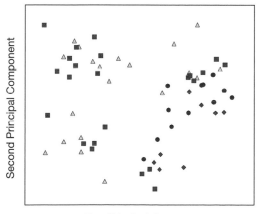

Figure 6.3   *Principal component analysis of 63 small, round blue cell tumors. Different symbols are used for each of the four categories as determined by classical diagnostics tests (see color insert).*

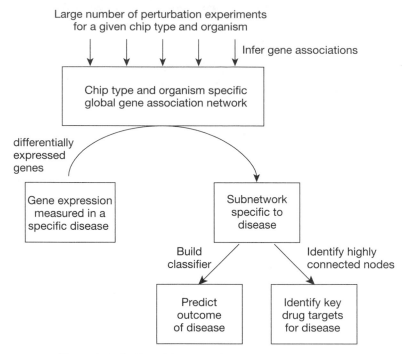

Figure 6.4 *Outline of the network approach to classification.*

There are two serious problems with this approach. As there are more than 20,000 human genes and typical DNA microarrays measure the activity of all of them, it is not possible to determine precisely the expression of all of them with just 100 experiments. The system is underdetermined. As a consequence, a resampling of 99 of the 100 patients will typically lead to a different selection of differentially expressed genes. In some cases there is little overlap between the different gene lists obtained by resampling, leading one to conclude that the gene selection is only slightly better than random because of the underdetermination of the entire system.

The other serious problem with the one gene–one function approach is that it ignores the network structure of gene expression. The expression of one gene is correlated to the expression of one or more other genes. Thus it is not useful to view genes as independent in this context. But when genes are selected using the *t*-test, such independence is exactly what is assumed. Instead of the one gene–one function model, a one network–many functions model is more useful and probably closer to the reality of molecular biology.

The structure of the gene network can be extracted from DNA microarray experiments. Two genes that are differentially expressed in the same experiment are connected either directly or indirectly. When such connected pairs are extracted from a large number of DNA microarray experiments, a substantial part of the gene network for a particular organism such as *Homo sapiens* can be deduced.

This network can be used to build better classifiers for a specific disease than those based on gene lists. Furthermore, the network can be used for making inferences about key drug targets for a specific disease (see Figure 6.4).

6.7.1 Construction of HG-U95Av2-Chip Based Gene Network

All HG-U95Av2 experiments from GEO (www.ncbi.nlm.nih.gov/geo) that fulfill the following criteria were selected and downloaded: single perturbation experiment where one cell or tissue type is compared under two or more comparable conditions. Cancer experiments were preferred as were experiments with raw CEL files. For the resulting 102 experiments all data files were logit normalized and connected genes identified with the *t*-test.

6.7.2 Construction of HG-U133A-Chip Based Gene Network

All HG-U133A experiments from GEO (www.ncbi.nlm.nih.gov/geo) that fulfill the following criteria were selected and downloaded: single perturbation experiment where one cell or tissue type is compared under two or more comparable conditions. Cancer experiments were preferred as were experiments with raw CEL files. For the resulting 160 experiments all data files were logit normalized and connected genes identified with the *t*-test. Genes observed as associated in at least two experiments were assigned to the gene network.

6.7.3 Example: Lung Cancer Classification with Mapped Subnetwork

As an example of the effect of using the mapped subnetwork, lung cancer patients (NSCLC adenocarcinoma) where classified with and without the mapped subnetwork (Figure 6.5). This dataset was not included in the construction of the gene network.

The raw data was downloaded as CEL files from www.genome.wi.mit.edu and separated into two groups: those patients with poor outcome (death or recurrence of disease) and those patients with good outcome (disease-free survival for the duration of observation). All CEL files (of chip type HG-U95Av2) were logit normalized in order to make them comparable. Genes that are correlated with outcome were selected using the *t*-test (logit-*t*, Lemon, *Genome Biol.* 2003; **4**(10):R67). The top 500 ranking genes were mapped to the HG-U95Av2 gene network described above. The genes that

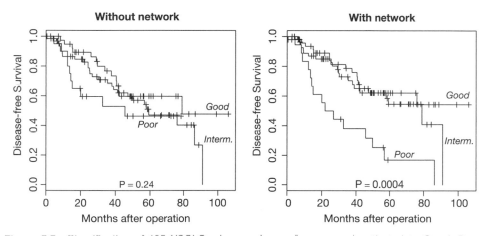

Figure 6.5 *Classification of 125 NSCLC adenocarcinoma (lung cancer) patients into Good, Poor, and Intermediate prognoses. The effect of using network mapping is shown. No other parameters are changed.*

mapped to the network were subjected to principal component analysis, and the 7 first principal components for each sample (patient) were retained.

Five different classification methods were trained on the principal components: *k* Nearest Neighbor (knn algorithm from www.r-project.org), Nearest Centroid, Support Vector Machine (svm algorithm from e1071 package at www.r-project.org), and Neural Network (nnet algorithm with 6 hidden units from nnet package at www.r-project.org). The classification was decided by voting among the five methods: Unanimous Good prognosis classification resulted in a Good prognosis prediction. Unanimous Poor prognosis classification resulted in a Poor prognosis prediction. Whenever there was disagreement between the methods, the Intermediate prognosis was predicted.

Testing of the performance of the classifier was done using leave-one-out cross-validation. One at a time, one patient (test sample) from one platform was left out of the gene selection and principal component selection as well as training of the five classifiers. Then the genes selected based on the remaining samples were extracted from the test sample and projected onto the principal components calculated based on the remaining samples. The resulting three principal components were input to five classifiers and used to predict the prognosis of the test sample. This entire procedure was repeated for all samples until a prediction had been obtained for all. The resulting prediction was plotted according to the clinical outcome (death or survival including censorship) in a Kaplan–Meier plot (Figure 6.5). Kaplan–Meier plots are further described in Chapter 7.

The effect of the network was determined by comparing the classification with the 500 genes before network mapping with the classification using only the mapped subnetwork.

Both the *P*-values in a log-rank test for difference in survival as well as the absolute difference in survival between the Good and Poor prognosis groups at the end of the experiment show a dramatic improvement with the network.

6.7.4   Example: Brain Cancer Classification with Mapped Subnetwork

The exact same procedure that was applied to the lung cancer dataset above was repeated with a HG-U95Av2-based brain cancer (glioma) dataset obtained from www.broad.wi.mit.edu/cancer/pub/glioma. The same HG-U95Av2 gene network was used. The resulting difference in outcome prediction with and without the mapped subnetwork is seen in Figure 6.6.

6.7.5   Example: Breast Cancer Classification with Mapped Subnetwork

The exact same procedure was applied to a breast cancer dataset downloaded from GEO (www.ncbi.nlm.nih.gov/geo, GEO accession number GSE2034). The HG-U133A gene network was used. The resulting difference in outcome prediction with and without the mapped subnetwork is seen in Figure 6.7.

6.7.6   Example: Leukemia Classification with Mapped Subnetwork

The exact same procedure was applied to acute myeloid leukemia using a dataset that was downloaded from www.stjuderesearch.org/data/AML1/index.html. The HG-U133A gene network was used. The resulting difference in outcome prediction with and without the mapped subnetwork is seen in Figure 6.8.

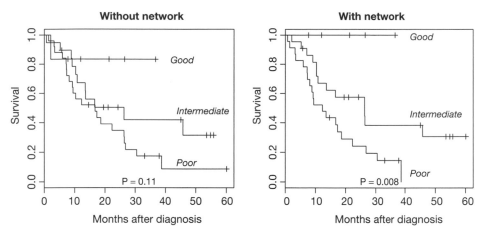

Figure 6.6 *Classification of 50 high-grade glioma (brain tumor) patients into Good, Poor, and Intermediate prognoses. The effect of using network mapping is shown. No other parameters are changed.*

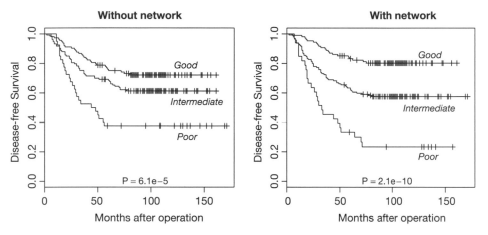

Figure 6.7 *Classification of 285 breast cancer patients into Good, Poor, and Intermediate prognoses. The effect of using network mapping is shown. No other parameters are changed.*

6.8 SUMMARY

The most important points in building a classifier are these:

- Collect as many examples as possible and divide them into a training set and a test set at random.
- Use as simple a classification method as possible with as few adjustable (learnable) parameters as possible. The k nearest neighbor method with 1 or 3 neighbors is a good choice that has performed well in many applications. Advanced methods (neural networks and support vector machines) require more examples for training than nearest neighbor methods.
- Test the performance of your classifier on the independent test set. This independent test set must not have been used for selection of features (genes). Average

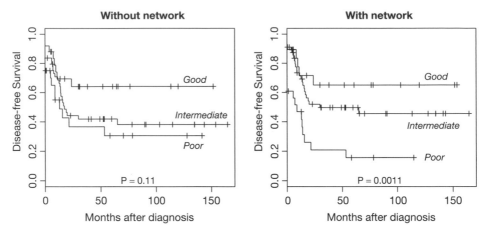

Figure 6.8 *Classification of 98 AML (leukemia) patients into Good, Poor, and Intermediate prognoses. The effect of using network mapping is shown. No other parameters are changed.*

the performance over several random divisions into training set and test set to get a more reliable estimate.

Can the performance on the independent test set be used to generalize to the performance on the whole population? Only if the independent test set is representative of the population as a whole and if the test set is large enough to minimize sampling errors.

A more detailed mathematical description of the classification methods mentioned in this chapter can be found in Dudoit et al. (2000a), which also contains other methods and procedures for building and optimizing a classifier.

FURTHER READING

Thykjaer, T., Workman, C., Kruhøffer, M., Demtröder, K., Wolf, H., Andersen, L. D., Frederiksen, C. M., Knudsen, S., and Ørntoft, T. F. (2001). Identification of gene expression patterns in superficial and invasive human bladder cancer. *Cancer Res.* 61:2492–2499.

Class Discovery and Classification

Antal, P., Fannes, G., Timmerman, D., Moreau, Y., and De Moor, B. (2003). Bayesian applications of belief networks and multilayer perceptrons for ovarian tumor classification with rejection. *Artif. Intell. Med.* 29(1-2):39–60.

Bicciato, S., Pandin, M., Didone, G., and Di Bello, C. (2003). Pattern identification and classification in gene expression data using an autoassociative neural network model. *Biotechnol. Bioeng.* 81(5):594–606.

Ghosh, D. (2002). Singular value decomposition regression models for classification of tumors from microarray experiments. *Pacific Symposium on Biocomputing* 2002:18–29. (Available online at http://psb.stanford.edu.)

Hastie, T., Tibshirani, R., Botstein, D., and Brown, P. (2001). Supervised harvesting of expression trees. *Genome Biol.* 2:RESEARCH0003 1–12.

Park, P. J., Pagano, M., and Bonetti, M. (2001). A nonparametric scoring algorithm for identifying informative genes from microarray Data. *Pacific Symposium on Biocomputing* 6:52–63. (Available online at http://psb.stanford.edu.)

Pochet, N., De Smet, F., Suykens, J. A., and De Moor, B. L. (2004). Systematic benchmarking of microarray data classification: assessing the role of nonlinearity and dimensionality reduction. *Bioinformatics* 20:3185–3195.

von Heydebreck, A., Huber, W., Poustka, A., and Vingron, M. (2001). Identifying splits with clear separation: a new class discovery method for gene expression data. *Bioinformatics* **17**(Suppl 1):S107–S114.

Xiong, M., Jin, L., Li, W., and Boerwinkle, E. (2000). Computational methods for gene expression-based tumor classification. *Biotechniques* 29:1264–1268.

Yeang, C. H., Ramaswamy, S., Tamayo, P., Mukherjee, S., Rifkin, R. M., Angelo, M., Reich, M., Lander, E., Mesirov, J., and Golub, T. (2001). Molecular classification of multiple tumor types. *Bioinformatics* 17(Suppl 1):S316–S322.

7

Survival Analysis

Survival analysis is concerned with the distribution of lifetimes. Often one is interested in how a certain treatment or diagnosis affects the survival of patients. Survival analysis offers a statistical framework for answering such questions.

The simplest form of survival analysis constitutes dividing the patients into two groups, one that receives the treatment and one that doesn't. Then the rate of survival at a given time point (e.g., five years after the start of the trial) between the two groups is compared and it is calculated whether there is a statistically significant difference.

7.1 KAPLAN–MEIER APPROACH

A more accurate approach is to compare survival curves. All patients might be dead after five years but in one group all patients might already be dead within the first two years. There could still be a significant difference, which is evident from the survival curve. The Kaplan–Meier approach of estimating survival curves even allows one to include patients who have not been observed for five years yet, but have been observed only for one year and are still alive. Such patients constitute *censored* data. At the time the observation stops (after one year), the patient is removed from the population of patients who are alive but is not added to the population of patients who have died.

Figure 7.1 shows survival curves for leukemia patients (Cox and Oakes, 1984). One group received the drug 6-mercaptopurine (6-MP) and the other group served as control. There is a clear difference in the survival curves for treatment and control. But is the difference significant? The *log-rank test* may be used to determine statistical significance. It gives us a P-value of 0.00004, which is highly significant.

Cancer Diagnostics with DNA Microarrays,   By Steen Knudsen
Copyright © 2006 John Wiley & Sons, Inc.

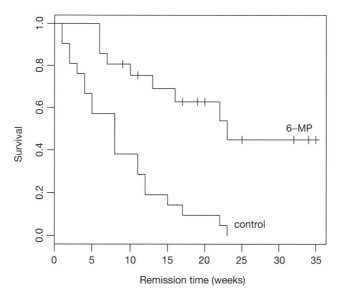

Figure 7.1 *Kaplan–Meier survival curves for leukemia patients divided into two groups. One group received treatment with 6-MP, the other group served as control. Crosses indicate time points where a patient has been removed due to censoring.*

7.2 COX PROPORTIONAL HAZARDS MODEL

The Cox proportional hazards model, on the other hand, fits a baseline survival curve to the data and then modifies it with proportional hazards, such as one or more treatments, or one or more genes with expression correlated to survival.

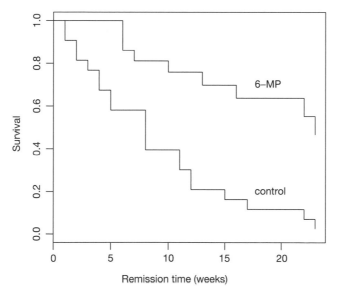

Figure 7.2 *Cox proportional hazards baseline survival curves for leukemia patients stratified into two groups. One group received treatment with 6-MP, the other group served as control.*

Figure 7.2 shows a Cox proportional hazards fit to each of the two groups of leukemia patients. Using a likelihood ratio test, the *P*-value is 0.00005, very similar to the log-rank test on the Kaplan–Meier curve.

7.3 SOFTWARE

The R code used to produce the plots and calculate the *P*-values shown in this chapter may be found in Section 10.5.3.

FURTHER READING

Venables, W. N., and Ripley, B. D. (1999). *Modern Applied Statistics with S-PLUS*, 3rd edition. New York: Springer.

Smith, P. J. (2002). *Analysis of failure and survival data*. Boca Raton: Chapman & Hall.

8

Meta-Analysis

We envisage a radical change in microarray studies—comparable to what happened in sequence analysis with the advent of the genome projects—where a division of labor takes place between a few large consortium-based projects on the one hand and the many smaller investigation-specific projects on the other hand.

—Moreau et al., 2003

In a meta-analysis, data from several microarray experiments on the same biological problem are combined in order to strengthen the conclusions. The experiments can be, and often are, performed on different microarray platforms.

The combining of several studies can be divided into two key tasks:

1. Match genes on one platform to genes on another platform.
2. Combine measurements for the same gene from several experiments.

There are several approaches to both tasks, and they are briefly mentioned in this chapter.

8.1 GENE MATCHING

The simplest way to match genes across platforms is to use the gene identifier associated with each gene on each platform. There are pitfalls to this approach, however. Each gene has several identifiers and an identity or similarity between identifiers may be missed unless some unified system is used. UniGene (http://www.ncbi.nlm.nih.gov/ entrez/query.fcgi?db=unigene) is one effort at creating such a unified system. RefSeq (http://www.ncbi.nlm.nih.gov/RefSeq/) is an even more reliable unified system.

Still, it has been shown that cross-platform consistency is significantly improved by sequence-based matching instead of identifier-based matching (Mecham et al., 2004). One of the many reasons why sequence-based matching is superior to gene identifier-based matching is that Affymetrix probes for one probeset may match several RefSeq genes (Gautier et al., 2004b).

8.2 COMBINING MEASUREMENTS

When a gene has been matched across platforms, there are several ways in which to combine the measurements from the different platforms.

8.2.1 Intensity

The intensity measurements of a gene matched across several platforms may be compared directly (Sorlie et al., 2003; Jiang et al., 2004; Mecham et al., 2004) and may be used to cluster or classify samples across datasets.

8.2.2 Effect Size

A standard way in statistical literature of combining studies in meta-analysis is through the combination of effect sizes (Choi et al., 2003, 2004). Effect sizes may be measured for each gene on each platform as the Z score that is the difference in means between two conditions divided by the pooled variance. These effect sizes may then be combined under a random-effects model (REM) or a fixed-effects model (FEM). The FEM assumes that differences in observed effect sizes are from sampling error alone. The REM assumes there are platform or study specific effects as well (Choi et al., 2003, 2004).

8.2.3 *P*-Values

An alternative approach is to apply significance testing separately on each platform and then combine the resulting P-values using Fisher's inverse chi-square method (Moreau et al., 2003). Wang et al. (2004) used a Bayesian approach to perform significance testing on combined measurements from different platforms.

8.2.4 Selected Features

It is also possible to select features from one dataset and use these features, if found on another platform, to classify samples on the other platform (Gruvberger et al., 2003; Sorlie et al., 2003).

8.2.5 Vote Counting

The way not to combine measurements is to apply significance testing separately on each platform and look for genes that are significant on all platforms. Because of the large number of false negatives expected with small number of replicates, this approach will be extremely inefficient and may find little or no overlap between the studies. It is preferable to rank genes according to how many platforms they are significant in. This is called vote counting.

8.3   META-CLASSIFICATION

A special case of meta-analysis is combining the results from several studies for a classifier of cancer diagnosis or cancer prognosis. Often this has been attempted by using the same genes to classify samples from different studies using different platforms. This approach has had limited success for the following reasons: (1) different platforms have different genes; (2) even if genes have been matched between platforms, they may be measuring different variants or splice forms; and (3) gene lists in themselves are ill-defined because it is not possible to establish with precision the expression of 10,000 genes or more in a single experiment. There will always be an element of randomness to the list of genes selected for a given classifier.

When the purpose of the microarray is to classify a tissue sample, as in cancer or no cancer, the exact identity of the genes that are useful for the classification becomes secondary. Because of the network structure of transcriptional regulation in human cells, there may be many genes that respond in a similar way—they are coexpressed.

So when combining measurements from different platforms it is not necessary to combine on a gene-by-gene basis. Instead, we can combine features that are essential to the classification problem. Such features can be regulatory pathways or groups of coexpressed genes. These features can be identified by a number of known mathematical methods applied to the matrix of expressed genes versus measured samples for each platform: principal component analysis (PCA), independent component analysis (ICA), correspondence analysis (CA), partial least squares (PLS), singular value decomposition (SVD), multidimensional scaling (MDS), and related methods.

What these mathematical methods have in common is that they project the large number of genes onto a smaller number of features, which capture essential information about the classification problem at hand. Furthermore, these derived features have become independent of the platform and can be combined between platforms.

So the classification problem is now posed as a number of features that characterize each sample from each platform. The first two features can be plotted on an $x-y$ coordinate system. Standard classification methods can now be applied to the samples represented by the features in a platform-independent way: K Nearest Neighbor, Nearest Centroid, Support Vector Machine, Neural Network, Linear Discriminant Analysis, and so on.

In the later chapters on individual cancer types (lung cancer, breast cancer, and lymphoma), examples of meta-classification using the above approach will be shown.

8.4   SUMMARY

When matching genes between platforms it is more reliable to match the sequence of the probes than to match the gene identifiers. Combining measurements from several platforms may be performed in many ways. But you should not compare lists of significant genes and look for overlap. With low numbers of replicates, the expectation is to find few genes in common even if the study group (patient population) is the same.

FURTHER READING

Best, C. J., Leiva, I. M., Chuaqui, R. F., Gillespie, J. W., Duray, P. H., Murgai, M., Zhao, Y., Simon, R., Kang, J. J., Green, J. E., Bostwick, D. G., Linehan, W. M., and Emmert-Buck,

M. R. (2003). Molecular differentiation of high- and moderate-grade human prostate cancer by cDNA microarray analysis. *Diagn. Mol. Pathol.* 12(2):63–70.

Ghosh, D., Barette, T. R., Rhodes, D., and Chinnaiyan, A. M. (2003). Statistical issues and methods for meta-analysis of microarray data: a case study in prostate cancer. *Funct. Integr. Genomics* 3(4):180–188.

Hedges, L. V., and Olkin, I. (1985). *Statistical Methods for Meta-Analysis*. New York: Academic Press.

Rhodes, D. R., Barrette, T. R., Rubin, M. A., Ghosh, D., and Chinnaiyan, A. M. (2002). Meta-analysis of microarrays: interstudy validation of gene expression profiles reveals pathway dysregulation in prostate cancer. *Cancer Res.* 62(15):4427–4433.

Rhodes, D. R., Yu, J., Shanker, K., Deshpande, N., Varambally, R., Ghosh, D., Barrette, T., Pandey, A., and Chinnaiyan, A. M. (2004). Large-scale meta-analysis of cancer microarray data identifies common transcriptional profiles of neoplastic transformation and progression. *Proc. Natl. Acad. Sci. USA* 101(25):9309–9314.

Sutton, A. J., Abrams, K. R., Jones, D. J., Sheldon, T. A. and Song, F. J. (2000). *Methods for Meta-Analysis in Medical Research*. Hoboken, NJ: Wiley.

9

The Design of Probes for Arrays

9.1 GENE FINDING

In the human genome there are a large fraction of genes that are not yet functionally characterized. They have been predicted either by the existence of a cDNA or EST clone with matching sequence, by a match to a homologous gene in another organism, or by *gene finding* in the genomic sequence. Gene finding uses computer software to predict the structure of genes based on DNA sequence alone (Guigo et al., 1992). Hopefully, they are marked as hypothetical genes by the annotator.

For certain purposes, for example, when designing a chip to measure all genes of a new microorganism, you may not be able to rely exclusively on functionally characterized genes and genes identified by homology. To get a better coverage of genes in the organism you may have to include those predicted by gene finding. Then it is important to judge the quality of the gene finding methods and approaches that have been used. While expression analysis may be considered a good method for experimental verification of predicted genes (if you find expression of the predicted gene it confirms the prediction), this method can become a costly verification if there are hundreds of false positive predictions that all have to be tested by synthesis of complementary oligonucleotides. A recent study showed that for *Escherichia coli* the predicted number of 4300 genes probably contains about 500 false positive predictions (Skovgaard et al., 2001). The most extreme case is the Archaea *Aeropyrum pernix*, where all open reading frames longer than 100 triplets were annotated as genes. Half of these predictions are probably false (Skovgaard et al., 2001).

Whether you are working with a prokaryote or a eukaryote, you can assess the quality of the gene finding by looking at which methods were used. If the only method used is looking for open reading frames as in the *A. pernix* case cited above, the worst prediction accuracy will result. Better performance is achieved when including codon usage (triplet) statistics or higher-order statistics (sixth-order statistics, e.g., measure frequencies of hexamers). These frequencies are to some degree specific to the organism (Cole et al., 1998). Even better performance is obtained when including models for specific signals like splice sites (Brunak et al., 1990a,b, 1991), promoters (Knudsen, 1999; Scherf et al., 2000), and start codons (Guigo et al., 1992). Such signals are best combined within hidden Markov models, which seem particularly well suited to the sequential nature of gene structure (Borodovski and McIninch, 1993; Burge and Karlin, 1997; Krogh, 1997).

9.2   SELECTION OF REGIONS WITHIN GENES

Once you have the list of genes you wish to spot on the array, the next question is one of cross-hybridization. How can you prevent spotting probes that are complementary to more than one gene? This question is of particular importance if you are working with a gene family with similarities in sequence. There is software available to help search for regions that have least similarity (determined by BLAST; Altschul et al., 1990) to other genes. At our lab we have developed ProbeWiz[1] (Nielsen and Knudsen, 2002), which takes a list of gene identifiers and uses BLAST to find regions in those genes that are the least homologous to other genes. It uses a database of the genome from the organism with which you are working. Current databases available include *Homo sapiens, Caenorhabditis elegans, Drosophila melanogaster, Arabidopsis thaliana, Saccharomyces cerevisae,* and *Escherichia coli.*

9.3   SELECTION OF PRIMERS FOR PCR

Once those unique regions have been identified, the probe needs to be designed from this region. It has been customary to design primers that can be used for polymerase chain reaction (PCR) amplification of a probe of desired length. ProbeWiz will suggest such primers if you tell it what length of the probe you prefer and whether you prefer to have the probe as close to the 3′ end of your mRNA as possible. It will attempt to select primers whose melting temperatures match as much as possible.

9.3.1   Example: Finding PCR Primers for Gene AF105374

GenBank accession number AF105374 (*Homo sapiens* heparan sulfate D-glucosaminyl 3-*O*-sulfotransferase-2) has been submitted to the web version of ProbeWiz, and the output generated if standard settings are used is given in Figure 9.1.

In addition to suggesting two primers for the PCR amplification, ProbeWiz gives detailed information on each of these primers and their properties, as well as a number of scoring results from the internal weighting process that went into selection of these two primers over others. The latter information may be useful only if you ask for more than one suggestion per gene and if you are comparing different suggestions.

---

[1] Available in a web version at http://www.cbs.dtu.dk/services/DNAarray/probewiz.html.

```
EST ID AF105374

Left primer sequence TGATGATAGATATTATAAGCGACGATG

Right primer sequence AAGTTGTTTTCAGAGACAGTGTTTTTC

PCR product size 327

Primer pair penalty 0.6575 (Primer3)
```

	left primer	right primer
Position	1484	1810
Length	27	27
TM	59.8	60.4
GC %	33.3	33.3
Self annealing	6.00	5.00
End stability	8.60	7.30

Penalties:	Weighted	Unweighted
Homology	0	0
Primer quality	65.75	0.657
3'endness	158	158

Figure 9.1 *Output of ProbeWiz server upon submission of the human gene AF105374 (Homo sapiens heparan sulfate D-glucosaminyl 3-O-sulfotransferase-2).*

9.4 SELECTION OF UNIQUE OLIGOMER PROBES

There is a trend in spotted arrays to improve the array production step by using long oligonucleotides (50 to 70 nucleotides) instead of PCR products. It is also possible to use multiple 25-nucleotide probes for each gene as done by Affymetrix. Li and Stormo (2001) have run their DNA oligo (50–70 nucleotides) prediction software on a number of complete genomes and made the resulting lists available online (http://ural.wustl.edu/˜lif/probe.pl).

We have developed a tool, OligoWiz (http://www.cbs.dtu.dk/services/OligoWiz), that will allow you to design a set of optimal probes (long or short oligos) for any organism for which you know the genome (Nielsen et al., 2003). In addition to selecting oligos that are unique to each gene, it also tries to assure that the melting temperatures of the oligos selected for an array are as close to each other as possible (ΔT_m score)—if necessary by varying the length of the oligo.

In addition, OligoWiz assesses other properties of probes such as their position in the gene (distance to the 3′ end, Position score) and the quality of the DNA sequence in the region where the probe is selected (GATC-only score) (Figure 9.2). The user

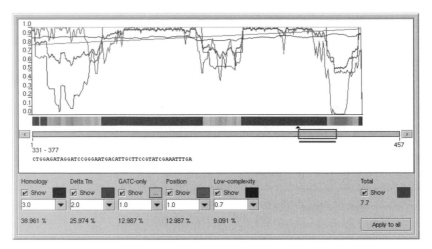

Figure 9.2 *Screenshot of part of the OligoWiz client program for designing oligonucleotide probes for a DNA microarray. The top window shows the distribution of the individual scores (defined in bottom window) over the entire gene.*

can assign a weight to the individual parameter scores associated with each probe that will affect the ranking of the probes that OligoWiz suggests for each gene.

9.5 REMAPPING OF PROBES

The assignment of probes to a gene is only as good as the bioinformatics used to design the probes. And the bioinformatics methods, as well as the annotation of the genomes based on which the probes are designed, improve all the time. For that reason a given probe-to-gene mapping can become outdated. For example, Affymetrix chips for human genes can be remapped using current human genome/transcriptome databases. We have performed such a remapping for some Affymetrix chips and have shown that it can affect the outcome of an analysis (Gautier et al., 2004). We have made the remapping available on our home page (http://www.cbs.dtu.dk/laurent/download/maprefseq/) together with the tools for remapping probes.

FURTHER READING

Primer and Oligo Probe Selection Tools

Carter, S. L., Eklund, A. C., Mecham, B. H., Kohane, I. S., and Szallasi, Z. (2005). Redefinition of Affymetrix probe sets by sequence overlap with cDNA microarray probes reduces cross-platform inconsistencies in cancer-associated gene expression measurements. *BMC Bioinformatics* **6**(1):107.

Varotto, C., Richly, E., Salamini, F., and Leister, D. (2001). GST-PRIME: a genome-wide primer design software for the generation of gene sequence tags. *Nucleic Acids Res.* **29**:4373–4377.

Rouillard, J. M., Herbert, C. J., and Zuker, M. (2002). OligoArray: genome-scale oligonucleotide design for microarrays. *Bioinformatics* **18**(3):486–487.

Gene Finding

Altschul, S. F., Madden, T. L., Schäffer, A. A., Zhang, J., Zhang, Z., Miller, W., and Lipman, D. J. (1997). Gapped BLAST and PSI-BLAST: a new generation of protein database search programs. *Nucleic Acids Res.* **25**:3389–3402. (Available at http://www.ncbi.nlm.nih.gov/BLAST/.)

## 10

Software Issues and Data Formats

In the future, sophisticated statistical, computational, and database methods may be as commonplace in Molecular Biology and Genetics as recombinant DNA is today.

—Pearson, 2001

Software for array data analysis is a difficult issue. There are commercial software solutions and there are noncommercial, public domain software solutions. In general, the commercial software lacks flexibility or sophistication and the noncommercial software lacks user-friendliness or stability.

Take Microsoft's Excel spreadsheet, for example. Many biologists use it for their array data processing, and indeed it can perform many of the statistical and numerical analysis methods described in this book. But there are pitfalls. First, commercial software packages can make assumptions about your data without asking you. As a scientist, you do not like to lose control over your calculations in that way. Second, large spreadsheets can become unwieldy and time-consuming to process and software stability can become an issue. Third, complicated operations require macroprogramming, and there are other, more flexible environments for programming.

You can perform the same types of analysis using noncommercial software. The best choice for microarray data analysis is the R statistics programming environment (www.r-project.org), where the Bioconductor consortium (www.bioconductor.org) has implemented most of the microarray analysis methods mentioned in this book. Numerous examples will be given later in this chapter. Extensive documentation on how to perform the analyses mentioned in this book are available from the Bioconductor web site.

There is, however, a steep learning curve. R is a very powerful language but to use it you must know about its objects and their structure. If this is not your cup of tea, there are other alternatives. You can use an automated, web-based approach such as GenePublisher (http://www.cbs.dtu.dk/services/GenePublisher), which is free and is based on Bioconductor. Or you can invest in a commercial software solution that integrates all analysis methods into one application.

10.1 STANDARDIZATION EFFORTS

There are several efforts underway to standardize file formats as well as the description of array data and underlying experiments. Such a standard would be useful for the construction of databases and for the exchange of data between databases and between software packages.

The Microarray Gene Expression Database Group (www.mged.org) has proposed Minimum Information About a Microarray Experiment (MIAME) for use in databases and in publishing results from microarray experiments.

MGED is also behind the MAGE-ML standard for exchanging microarray data. MAGE-ML is based on the XML standard used on the World Wide Web. It uses tags, known from HTML, to describe array data and information.

10.2 DATABASES

A number of public repositories of microarray data have emerged. Among the larger ones are:

- *ArrayExpress*
  European Bioinformatics Institute
  (http://www.ebi.ac.uk/arrayexpress/)
- *Gene Expression Omnibus*
  National Institutes of Health
  (http://www.ncbi.nlm.nih.gov/geo/)
- *Stanford Microarray Database*
  Stanford University
  (http://genome-www5.stanford.edu/)

10.3 STANDARD FILE FORMAT

The following standard file format is convenient for handling and analyzing expression data and will work with most software (including the web-based GenePublisher). Information about the experiment, the array technology, and the chip layout is not included in this format and should be supplied in separate files. The file should be one line of tab-delimited fields for each gene (probe set):

Field 1: Gene ID or GenBank accession number

Field 2: (Optional) text describing function of gene

Fields 3 and up: Intensity values for each experiment including control, or logfold change for each experiment relative to the control

The file should be in text format, and not in proprietary, inaccessible formats that some commercial software houses are fond of using.

10.4 SOFTWARE FOR CLUSTERING

- *Cluster*
  A widely used and user friendly software by Michael Eisen (http://rana.lbl.gov) It is for Microsoft Windows only.
- *GeneCluster*
  From the Whitehead/MIT Genome Center. Available for PC, Mac, and Unix (http://www-genome.wi.mit.edu/MPR/)
- *Expression Profiler*
  A set of tools from the European Bioinformatics Institute that perform clustering, pattern discovery, and gene ontology browsing. Runs in a web server version and is also available for download in a Linux version (http://ep.ebi.ac.uk)

10.5 SOFTWARE FOR STATISTICAL ANALYSIS

Microsoft Excel has several statistics functions built in, but an even better choice is the free, public-domain R package, which can be downloaded for Unix/Linux, Mac, and Windows systems (http://www.r-project.org). This software can be used for t-test, ANOVA, principal component analysis, clustering, classification, neural networks, and much more. The Bioconductor (http://www.bioconductor.org) consortium has implemented most of the array-specific normalization and statistics methods described in this book.

10.5.1 Example: Statistical Analysis with R

In this section we use a small standard example for clarity (Table 10.1). There are four genes, each measured in six patients, which fall into three categories: normal (N), disease stage A, and disease stage B. That means that each category has been *replicated* once. The data follow the standard format except that there is no function description.

Next you boot up R (by typing the letter R at the prompt) and read in the file:

```
dataf  <- read.table("example")
```

In the web companion to this book you can find the example and code for copy-paste to your own computer (http://www.cbs.dtu.dk/staff/steen/book.html).

TABLE 10.1 Expression Readings of Four Genes in Six Patients

Gene	Patient					
	N_1	N_2	A_1	A_2	B_1	B_2
a	90	110	190	210	290	310
b	190	210	390	410	590	610
c	90	110	110	90	120	80
d	200	100	400	90	600	200

10.5.1.1 The t-test Here is how you would perform a *t*-test to see if genes differ significantly between patient category A and patient category B:

```
# load library for t-test:
  library(ctest)
# t-test function:
  get.pval.ttest <- function(dataf,index1,index2,
    datafilter=as.numeric){
    f <- function(i) {
      return(t.test(datafilter(dataf[i,index1]),
        datafilter(dataf[i,index2]))$p.value)
    }
    return(sapply(1:length(dataf[,1]),f))
  }
# call function with our data:
  pVal.ttest <- get.pval.ttest(dataf,3:4,5:6)
# print results on screen (only for a small dataset
                                       like this)
  print(cbind(dataf,pVal.ttest))
        V1  V2  V3  V4  V5  V6  V7 pVal.ttest
  1 gene_a  90 110 190 210 290 310 0.01941932
  2 gene_b 190 210 390 410 590 610 0.00496281
  3 gene_c  90 110 110  90 120  80 1.00000000
  4 gene_d 200 100 400  90 600 200 0.60590011
# sort on P-value and write to file:
  orders <- order(pVal.ttest)
  ordered.data <- cbind(dataf[orders,],pVal.ttest[orders])
  write(t(as.matrix(ordered.data)),
   ncolumns=length(dataf)+1,
   file = "ttest.out")
  q(save="no")
```

The default call to the `t.test` function assumes unequal variance between the two populations (Welch's *t*-test).

10.5.1.2 Wilcoxon You can perform the Wilcoxon test instead of the *t*-test by replacing the call to the `t.test` function above with a call to the `wilcox.test` function.

10.5.1.3 ANOVA Here is how you would perform an ANOVA on the example to test for genes that differ significantly in at least one of the three categories N, A, and B:

```
# Specify categories and columns holding AvgDiff data:
  Categories <- as.factor(c("O","O","A","A","B","B"))
  indexAvgDiff <- 1:6
# ANOVA function:
  get.pval.anova <-
  function(dataf,indexAll,Categories,
    datafilter=as.numeric){
    Categories <- as.factor(Categories)
    f <- function(i) {
      return(summary(
          aov(datafilter(dataf[i,indexAll]) ~ Categories)
        )
      [[1]][4:5][[2]][1])
```

```
    }
    return(sapply(1:length(dataf[,1]),f))
  }
# call the function with our data:
  pVal.anova <- get.pval.anova(dataf,indexAvgDiff,
                               Categories)
# print results on screen (only for a small dataset
                                      like this)
  print(cbind(dataf,pVal.anova))
          V1  V2  V3  V4  V5  V6  V7   pVal.anova
  1 gene_a  90 110 190 210 290 310 0.0017965439
  2 gene_b 190 210 390 410 590 610 0.0002283540
  3 gene_c  90 110 110  90 120  80 1.0000000000
  4 gene_d 200 100 400  90 600 200 0.5560965577
# sort on P-value and write results to file:
  orders <- order(pVal.anova)
  ordered.data <- cbind(dataf[orders,],pVal.anova[orders])
  write(t(as.matrix(ordered.data)),
   ncolumns=length(dataf)+1,
   file = "ANOVA.out")
  q(save="no")
```

In general, it is best to perform statistical tests on the raw expression values instead of fold change. If you are working with spotted arrays where there is much variation between slides, it may be better to perform the statistical test on the fold change derived from the red and green channels of each slide. Try to perform the statistical test on both absolute values and fold change and compare the results. Baldi and Long (2001) advocate log-transformation of data before statistical analysis.

10.5.1.4 PCA   You can perform a principal component analysis of the same data as follows (note that the first row of the data must contain labels of all patients, but no label for the gene identifier):

```
library(mva)
dataf  <- read.table("example")
pca <- princomp(dataf)
summary(pca)
plot(pca)
biplot(pca)
```

This will produce the plot shown in Figure 4.2 in Section 4.1. The plot shown in Figure 4.3 was produced by transposing the data matrix with the command t(dataf), but PCA may not work for large transposed matrices.

10.5.1.5 Correspondence Analysis   You can perform a correspondence analysis much the same way as you perform a principal component analysis:

```
library(MASS)
library(mva)
dataf  <- read.table("example",header=T)
cs <- corresp(dataf,nf=2)
plot(cs)
```

10.5.1.6 Classification You can perform *k* nearest neighbor classification with cross-validation using built-in functions in the standard distribution of R. This is the calculation that was performed for the classifier in Section 6.6:

```
library(class)
library(mva)
dataf  <- read.table("datafile",header=T)
tposed <- t(dataf)
knn.targets <- factor( c(rep("E", 23), rep("B", 8),
              rep("N", 12), rep("R", 20)) )
knn.cv(tposed, knn.targets, k=3, prob=TRUE)
E E E E E E E E E E E E E E E E E E E E E E E
B B B B B B B B N N N N N N N N N N N R R R N
R R R R R E R R R R R R R R R
```

The predicted classes (last two lines) differ from those assigned (knn.targets) for two tumors, so this classification is 97% correct.

10.5.1.7 Neural Networks The R package has a function for training a feed-forward neural network and using the trained network to predict the class of unlabeled samples. These calculations were performed for the neural network results shown in Section 6.6 with leave-one-out cross-validation:

```
library(class)
library(nnet)
pca <- princomp(tposed,cor=TRUE,scores=TRUE)
canc <- pca$scores[,1:10]
targets <- class.ind( c(rep("E", 23), rep("B", 8),
                rep("N", 12), rep("R", 20)) )
f <- function(samp) {
    b <- c(0,0,0,0)
    for(i in 1:20) {
        trainednet <- nnet(canc[-samp,], targets[-samp,],
              size=2,skip=FALSE,trace=FALSE, maxit=300)
        a <- max.col(predict.nnet(trainednet, canc[samp,]))
        b[a] <- b[a] + 1
    }
print(b)
}
lapply(1:63,f)
```

10.5.2 The affy Package of Bioconductor

We have contributed to the affy package (Gautier et al., 2004a) at Bioconductor, which, in addition to the above-mentioned statistical analyses, can replace Affymetrix GeneChip software by reading CEL files directly and calculating expression values using the Li–Wong method as well as other methods. It can normalize the data using a range of signal-dependent normalization methods. Here is an example of how the functions in the affy package can be called to read and analyze a batch of Affymetrix CEL files (start by loading R version 1.7.1 or higher):

```
# use version 1.2.30 or higher of the affy package:
  library(affy)
# read CEL files:
  data <- read.affybatch("day_7a_amp.CEL", "day_7b_amp.CEL",
         "day_7c_amp.CEL", "day_7a_HIV_amp.CEL",
         "day_7b_HIV_amp.CEL", "day_7c_HIV_amp.CEL")
 # background correct, normalize, and condense:
  affy.es <- expresso(data, bgcorrect.method="rma2",
         normalize.method = "qspline",
         pmcorrect.method="pmonly",
         summary.method ="liwong")
  Matrix <- exprs(affy.es)
 # t-test function:
   get.pval.ttest <- function(dataf,index1,index2,
         datafilter=as.numeric)
         f <- function(i)
            return(t.test(datafilter(dataf[i,index1]),
              datafilter(dataf[i,index2]))$p.value)

         return(sapply(1:length(dataf[,1]),f))

 # perform t-test:
  pValues <- get.pval.ttest(Matrix,1:3,4:6)
 # write ranked results to file:
  orders <- order(pValues)
  ordered.data <- cbind(rownames(Matrix)[orders],
                 Matrix[orders,],pValues[orders])
  write(t(as.matrix(ordered.data)),
        ncolumns=1+ncol(Matrix)+1,file = "Pvalues.abs")
```

The input data files are available in the web companion for this book (http://www.cbs.dtu.dk/staff/steen/book.html).

Counterintuitively, the affy package can be used to analyze data from spotted arrays and other platforms as well. Here is an example using an input file in the standard tab-delimited text file format described above with no annotation field and no header (start by loading R version 1.7.1 or higher):

```
  library(affy)
# read file:
  dataf <- read.table("input.file", row.names=1, sep="",
               header=F, comment.char = "")
# normalize using qspline:
  normdata <- normalize.qspline(dataf, na.rm=TRUE)
 # t-test function:
   get.pval.ttest <- function(dataf,index1,index2,
         datafilter=as.numeric)
         f <- function(i)
            return(t.test(datafilter(dataf[i,index1]),
              datafilter(dataf[i,index2]))$p.value)

         return(sapply(1:length(dataf[,1]),f))

 # perform t-test:
  pValues <- get.pval.ttest(normdata,1:12,13:24)
```

```
# write ranked results to file:
  orders <- order(pValues)
  ordered.data <- cbind(rownames(dataf)[orders],
          normdata[orders,],pValues[orders])
  write(t(as.matrix(ordered.data)),
      ncolumns=1+ncol(normdata)+1,file = "Pvalues.abs")
```

10.5.3 Survival Analysis

10.5.3.1 Kaplan–Meier Using the library *survival* of the R project it is possible to calculate survival curves:

```
# plot survival curves:
  library(survival)
  data(gehan)
  gehan.surv <- survfit(Surv(time, cens)    treat, data
    = gehan, conf.type = "log-log")
  plot(gehan.surv, ylab="Survival", xlab="Remission time
    (weeks)")
  text(30,0.5,pos=4,"6-MP"); text(23,0.05,pos=4, "control")
 # perform log-rank test for significance of difference
    between curves:
  survdiff(Surv(time, cens)    treat, data = gehan)
```

10.5.3.2 Cox Proportional Hazards A Cox model on the same data can be fitted as follows:

```
fit <- coxph(Surv(time, cens)    strata(treat), data=gehan)
plot(survfit(fit), ylab="Survival", xlab="Remission time
    (weeks)")
text(17,0.7,pos=4, "6-MP"); text(17,0.2,pos=4, "control")
```

10.5.4 Commercial Statistics Packages

The commercial statistics packages SAS, SPSS, and S-PLUS include the statistical functions described in this section as well.

10.6 SUMMARY

The R programming language with the Bioconductor packages offer the best solution for data analysis. They run on any PC, Mac, or Unix system. With such a setup your possibilities for data mining and discovery are limited only by your imagination (and, perhaps, by your programming skills).

FURTHER READING

Parmigiani, G., Garrett, E. S., Irizarry, R. A., and Zeger, S. (Editors). (2003). *The Analysis of Gene Expression Data.* Berlin: Springer Verlag.

R

Online manuals available at http://www.r-project.org.

Venables, W. N. and Ripley, B. D. (1999). *Modern Applied Statistics with S-PLUS*, 3rd ed. New York: Springer.

11

Breast Cancer

11.1   INTRODUCTION

Breast cancer is the second leading cause of cancer death among women, exceeded only by lung cancer. In 2006, about 40,970 women and 460 men will die from breast cancer in the United States (American Cancer Society, 2006). More than 212,000 women will be diagnosed with invasive (stages I–IV) breast cancer.

11.1.1   Anatomy of the Breast

The female breast (Figure 11.1) consists of milk glands (lobules), ducts that connect the glands to the nipple, and stroma (adipose tissue, connective tissue, lymph vessels and blood vessels). Cancers can originate in all of these tissues, but most of them originate in the lobules.

11.1.2   Breast Tumors

Fibroadenomas and papillomas constitute tumors that cannot spread to other organs and are thus considered benign. They constitute the majority of lumps discovered in the breast. Cancers originating in the ducts and lobules are categorized as adenocarcinomas. They are divided into those that have not spread from the originating tissue (referred to as *carcinoma in situ*, or CIS), and those that have infiltrated or invaded the surrounding tissue (referred to as IC). So common breast cancers are lobular carcinoma in situ (LCIS), ductal carcinoma in situ (DCIS), infiltrating lobular carcinoma (ILC), and infiltrating ductal carcinoma (IDC).

More rare types of breast cancer include inflammatory breast cancer, medullary carcinoma, mucinous carcinoma, Paget's disease of the nipple, Phyllodes tumor, and tubular carcinoma.

Cancer Diagnostics with DNA Microarrays,   By Steen Knudsen
Copyright © 2006 John Wiley & Sons, Inc.

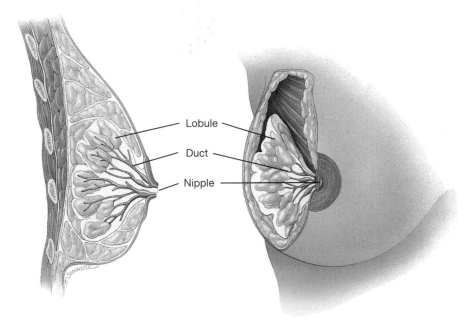

Figure 11.1 Anatomy of the female breast. (From Tortora, Principles of Human Anatomy, 10th ed., 2005, p. 865. Used with permission of John Wiley & Sons, Inc.)

11.2 CURRENT DIAGNOSIS

Current detection of breast cancer is by mammography and physical examination for lumps. Those that are positive in these initial population-wide screens are then further examined by, for example, ultrasound, biopsies, magnetic resonance imaging (MRI), and positron emission tomography (PET).

11.2.1 Staging

A comprehensive examination of the breast cancer patient, including surgery, allows the division of the cancer progression into stages that have different prognoses. The staging is based on the tumor size (T) in centimeters, on whether the cancer has spread to nearby (axillary) lymph nodes (N), and whether or not metastasis to other parts of the body has been observed (M).

Stage 0   Carcinoma in situ. No infiltration of surrounding tissue, no spreading to lymph nodes or distant sites (metastasis).

Stage I   The tumor is less than 2 cm in diameter and has not spread to lymph nodes or distant sites (metastasis).

Stage IIA   Either there is no tumor found but it has spread to 1–3 axillary lymph nodes, or the tumor is less than 2 cm in diameter and has spread to 1–3 axillary lymph nodes, or the tumor is between 2 and 5 cm in diameter but has not spread to lymph nodes. There is no metastasis to distant sites.

Stage IIB   The tumor is between 2 and 5 cm and is found in 1–3 axillary lymph nodes, or the tumor is larger than 5 cm and does not grow into the chest wall and has not spread to the lymph nodes. There is no metastasis.

Stage IIIA   The tumor is smaller than 5 cm and has spread to 4–9 axillary lymph nodes, or the tumor is larger than 5 cm and has spread to 1–9 axillary nodes.

Stage IIIB   The tumor grows into the chest wall or skin and has spread to 0–9 axillary lymph nodes. There is no metastasis.

Stage IIIC   The tumor is any size, has spread to 10 or more axillary lymph nodes. There is no metastasis.

Stage IV   The cancer, regardless of its size, has spread to distant organs such as bone, liver, or lung or to lymph nodes far from the breast.

An alternative to axillary lymph node detection is sentinal lymph node detection in the breast. For a full description of the staging system refer to the 2002 American Joint Committee on Cancer. The five-year relative survival rates (excluding deaths from other causes) have been calculated retrospectively for the individual stages of breast cancer (American Cancer Society, 2003). They are shown in Table 11.1.

11.2.2   Histopathological Grading

Tumor tissue removed during surgery or biopsy is examined by a pathologist and graded according to the differentiation of the cells. Grade 1 is well-differentiated like normal cells; grade 3 is poorly differentiated—a hallmark of cancer cells. Grade 2 is in between grade 1 and grade 3.

11.2.3   Clinical Markers

Receptors for the hormones estrogen and progesterone play a role in the control of growth of the cell. Tumor cells often lose these normal function of receptors, allowing the cells to grow uncontrolled. Immunoassays are used to establish the estrogen receptor (ER) and progesterone receptor (PR) status of the removed tumor cells. ER positive and PR positive patients have a better prognosis than ER negative and PR negative.

The *oncogene* HER2/neu is found overexpressed in rapidly spreading tumors. Immunohistochemical tests exist for the HER2/neu protein, and the copy number of the HER2/neu *gene* can be established with fluorescent in situ hybridization (FISH). The drug Herceptin, a monoclonal antibody against HER2/neu, has been developed to intervene in the function of the HER2/neu protein and has shown remarkable results in the treatment of HER2/neu overexpressing breast cancer patients.

TABLE 11.1   Five-Year Relative Survival Rates for Breast Cancer Patients According to Stage of the Disease

Stage	Five-Year Survival
0	100%
I	98%
IIA	88%
IIB	76%
IIIA	56%
IIIB	49%
IV	16%

11.3 CURRENT THERAPY

11.3.1 Surgery

The American Cancer Society together with the National Comprehensive Cancer Network provides guidelines for breast cancer treatment. Consensus conferences in the United States and in Europe also publish guidelines for the eligibility of adjuvant chemotherapy (St. Gallen and NIH consensus published in Goldhirsch et al. (1998) and Eifel et al. (2001)). The Nottingham Prognostic Index predicts the survival of patients after surgery based on pathological factors (Blamey et al., 1979). The tumor is removed by surgery, possibly followed by local radiation therapy. Neoadjuvant chemotherapy can be given before surgery to reduce the size of the tumor. Adjuvant chemotherapy after surgery is usually given as a combination of several of the following drugs: cyclophosphamide, methotrexate, fluorouracil, doxorubicin, paclitaxel (Taxol), and epirubicin.

11.3.2 Chemotherapy

The chemotherapy has side effects including fatigue, nausea, vomiting, hair loss, increased chance of infection, bleeding, premature menopause, infertility, heart disease, leukemia, and decreased mental functioning.

11.3.3 Hormone Therapy

Tamoxifen is an antagonist of the hormone estrogen, which is necessary for the growth of many tumors. Tamoxifen has been shown to increase survival. Long-term use of Tamoxifen (more than five years) has been shown to lead to increased risk of endometrial cancer (cancer of the lining of the uterus) and uterine sarcoma. Tamoxifen has also been associated with an increased risk of blood clots. Other drugs that reduce the effect of estrogen in some ways are also in testing or have been approved for treatment.

11.3.4 Monoclonal Antibody Therapy

Trastuzumab (Herceptin) is a monoclonal antibody that is specific to the growth-promoting HER2/neu protein. Herceptin can shrink some breast cancer metastases.

11.4 MICROARRAY STUDIES OF BREAST CANCER

11.4.1 The Stanford Group

The first application of DNA microarrays to the study of breast cancer was published by Perou and co-workers in *PNAS* in 1999. It studied 13 tumors using cDNA microarrays for 5531 human genes. It was quickly followed by a more elaborate study in *Nature* in 2000, using a spotted cDNA array consisting of 8102 human genes and analyzing 65 dissected tissues from 42 individuals. These included 36 infiltrating ductal carcinomas,

2 lobular carcinomas, 1 ductal carcinoma in situ, 1 fibroadenoma, and 3 normal breast samples.

From the genes they selected an "intrinsic" set of 496 genes that varied more between tumors than within repeated samplings of the same tumor, and they performed a hierarchical clustering of the samples based on these 496 genes. The sample clustered into two groups, which coincided with their estrogen receptor status. One cluster was ER positive, the other ER negative. The ER negative cluster contained at least two biologically distinct subtypes of tumor.

One of the surprising findings was that a metastasis and the primary tumor were as similar in the overall pattern of gene expression as were repeated samplings of the same tumor. This suggests that the genetic program of the primary tumor is generally retained in its metastases.

The study was later expanded to 85 tissue samples (Sørlie et al., 2001). These included 71 ductal carcinomas, 5 lobular carcinomas, 2 ductal carcinomas in situ, 3 fibroadenomas, and 5 normal breast samples. All but three patients were treated with Tamoxifen. From the 1753 genes that varied more than fourfold in at least three samples, a set of genes that were correlated to patient survival was selected. The SAM software (Tusher et al., 2001) was used to calculate a maximum-likelihood score statistic from Cox's proportional hazards model. A set of 264 survival-correlated genes was selected, and when a hierarchical clustering was performed based on these genes, the patients clustered in largely the same groups as those identified with the "intrinsic" set of 496 genes. Thus Kaplan–Meier survival curves were drawn for each of six subclasses identified by clustering. They showed a statistically significant difference in survival between the subtypes. Mutations in the *TP53* gene were found in DNA from patients in all subclasses, but with a different frequency.

Finally, in 2003, the group published an updated study with 115 samples (Sørlie, 2003) and compared it to the studies of van't Veer (see Section 11.4.2) and West (see Section 11.4.3). They identified genes that could be used to cluster all datasets and observed a remarkable agreement in the clustering, even though the West study was performed on a different platform, the Affymetrix GeneChip. There was also a remarkable agreement in outcome between the Sørlie and van't Veer studies. No outcome data was available for the West study at that time.

11.4.2 The Netherlands Cancer Institute Group

Van't Veer and co-workers published their first study in *Nature* in 2002. It included tumor samples from 117 young patients, including 34 patients who developed distant metastases within five years, 44 patients who remained disease-free after five years, 18 patients with germline *BRCA1* mutations, and 2 patients with *BRCA2* mutations. The Agilent inkjet platform was used with 25,000 human genes. Of these, 231 genes were correlated to clinical outcome (correlation coefficient smaller than -0.3 or larger than 0.3). Van't Veer and co-workers committed the classical mistake of using the test set to select the genes to be used in the classifier. This means that they fitted their classifier to their test set and overestimated its performance. An independent test set should have been used to evaluate the performance of the classifier. They then further reduced the set of genes by leave-one-out cross-validation to find the subset that gave the best performance on the test set. This left 70 genes. Here again, an independent test set would have been necessary to evaluate the performance of the classifier with the optimal set of genes.

A verification study was published the same year in the *New England Journal of Medicine* (van de Vijver et al., 2002). It consisted of 295 patients with stage I or stage II breast cancer and applied the 70-gene prognostic profile obtained in the *Nature* study. Sixty-one patients from their previous study were included in the verification study, however, meaning that it was not truly independent. They concluded that the 70-gene profile is a more powerful predictor of outcome than standard systems based on clinical and histological criteria. The authors have filed a patent on using gene expression profiling for breast cancer prognosis and have formed a company, Agendia Inc., to commercialize the patent.

11.4.3   The Duke University Group

A group from Duke University published a study based on the Affymetrix HuGeneFL GeneChip in *PNAS* in 2001 (West et al., 2001). The study used 49 tumors of stage II and above. They included 13 ER+ N+ tumors, 12 ER− N+ tumors, 12 ER+ N− tumors, and 12 ER− N− tumors. The study used binary regression models combined with singular value decompositions of the genes. This yielded a set of "metagenes" that could be used to predict ER status and lymph node status.

In 2003 the same group published a study in *The Lancet* of 89 tumor samples run on the Affymetrix U95Av2 GeneChip (Huang et al., 2003). The data analysis was again singular value decomposition to form a number of "metagenes" that were used in a Bayesian classification tree analysis. The resulting classifier was cross-validated and showed a 90% accuracy in predicting the outcome for an individual patient.

Finally, in 2004, the group (Pittman et al., 2004) published an expanded study of 158 Taiwanese breast cancer patients. It was still based on the Affymetrix U95Av2 GeneChip, but now clinical data was integrated into the Bayesian classification tree models. The resulting models are better at predicting clinical outcome (with 90% specificity and 90% sensitivity), which the authors claim is better than similar models based on expression data alone.

11.4.4   The Lund Group

The Lund University group, together with NIH, in 2001 published a study showing that the estrogen receptor status could be predicted based on microarray data and analyzed the genes involved (Gruvberger et al., 2001). They used 58 node-negative breast carcinomas and spotted cDNA arrays with 6728 clones. The analysis was based on PCA to reduce the dimensionality of the input genes. Principal components were used as input to a neural network trained to predict ER status.

In a later paper (Gruvberger et al., 2003), the group applied the van't Veer outcome predictor profile to their data. Fifty-eight out of the 231 outcome-correlated genes identified by van't Veer and co-workers were also found on the arrays used by Gruvberger and co-workers. They were, however, not significantly correlated to outcome. Gruvberger and co-workers were not able to identify other genes significantly correlated with outcome, and they explain this discrepancy to the van't Veer data in that the van't Veer cohort has a high correlation between outcome and ER status. Thus the genes that predict ER status overlap the genes that predict outcome. So they suggest that the van't Veer outcome predictor may be a modified predictor of ER status that can predict outcome in cohorts where ER status and outcome are correlated but not necessarily in other cohorts such as that used by Gruvberger. The cohort of Gruvberger was selected to consist of four nearly equal-sized groups: ER positive and ER negative, each divided into Good and Poor prognosis groups, thus eliminating any obvious correlation between ER status and prognosis in the cohort. Some of the patients received hormone treatment after surgery.

Finally, the group published a paper (Eden et al., 2004) to show that the prognostic accuracy obtained by van't Veer and co-workers in the van't Veer cohort does not exceed the prognostic accuracy of a weighted sum of conventional clinical markers.

11.4.5 The Karolinska Group

Pawitan et al. (2005) published a study of 159 patients using Affymetrix U133A and U133B GeneChips. Some of the patients had received adjuvant chemotherapy. A subset of 64 genes was found to be correlated to outcome. Hierarchical clustering revealed three subgroups: patients who did well with therapy, patients who did well without therapy, and patients who failed to benefit from given therapy. Independent validation was performed on a group of 211 patients. Prognosis was more accurate than conventional markers when analyzed individually.

11.4.6 The Tokyo Group

Nagahata et al. (2004) used a 25,344-clone cDNA array to study 20 breast cancer patients, half of whom had died within five years of surgery and half of whom had survived disease-free for more than five years after surgery. They identified 71 genes that were correlated to the outcome of the disease.

11.4.7 The Veridex Group

Y. Wang et al. (2005) published a large study of 286 lymph node negative breast cancer patients using Affymetrix HG-U133A GeneChips. They found a convincing ability to differentiate the prognoses in this group of patients who are already predicted by conventional staging to have a Good prognosis. The classification worked for estrogen receptor positive patients as well as for estrogen receptor negative patients. The data has been deposited at GEO (GSE2034).

11.4.8 The National Cancer Institute Group

Sotiriou et al. (2003) published a study of 99 primary breast carcinomas analyzed on a 7650-feature cDNA array. Cox proportional hazards regression analysis identified 16 genes that were significantly associated with relapse free survival.

11.5 META-CLASSIFICATION OF BREAST CANCER

Five of the studies listed above (Stanford, Netherlands, Duke, Veridex, and Sotiriou) have made their raw data available, allowing a meta-classification where one classifier classifies all patients from all studies based on their principal components.

The raw data was logit-normalized and genes that correlated with outcome were selected by t-test for each study. The 100 top ranking genes in this test were subjected

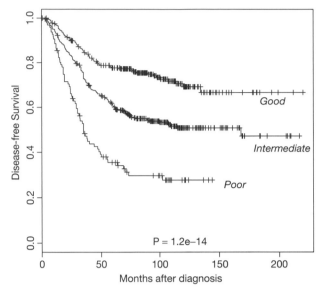

Figure 11.2   *Survival of 845 breast cancer patients combined from five different studies. The patients are grouped according to the DNA chip predicted prognosis (Good, Intermediate, Poor) based on the MetaClassifier tool using principal component analysis (PCA).*

to principal component analysis for each study, and the 3 first principal components were retained (using 500 genes and 10 components gives almost the same result).

The principal components from all five datasets were combined as shown in Figure 11.2 and were used to develop a classifier to predict outcome of the disease of the patient. First the arbitrary signs of the principal components from each platform had to be adjusted. For each principal component we chose the sign that made the sum of squared errors over classes and datasets smallest.

Four different classification methods were trained on the principal components: *k* Nearest Neighbor (knn algorithm from www.r-project.org), Nearest Centroid, Support Vector Machine (svm algorithm from e1071 package at www.r-project.org), and Neural Network (nnet algorithm with 6 hidden units from nnet package at www.r-project.org). The classification was decided by voting among the four methods: Unanimous Good prognosis classification in a Good prognosis prediction. Unanimous Poor prognosis classification resulted in a Poor prognosis prediction. Whenever there was disagreement between the methods, the Intermediate prognosis was predicted.

Testing of the performance of the classifier was done using leave-one-out cross-validation. One at a time, one patient (test sample) from one platform was left out of the gene selection, principal component selection, and sign adjustment as well as training of the five classifiers. Then the genes selected based on the remaining samples were extracted from the test sample and projected onto the principal components calculated based on the remaining samples. The resulting three principal components were input to five classifiers and used to predict the prognosis of the test sample. This entire procedure was repeated for all samples until a prediction had been obtained for all.

The performance of the classifier on the total of 845 patients was tested by leave-one-out cross-validation and the results are shown in a Kaplan–Meier survival plot seen in Figure 11.2. Showing only the 339 estrogen receptor positive and lymph node

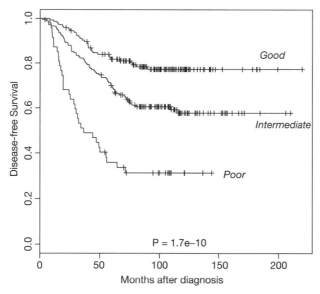

Figure 11.3 *Survival of 369 estrogen receptor positive (ER+), lymph node negative (LN−) breast cancer patients combined from four studies where the ER and LN clinical findings are available. These patients are considered to have a Good prognosis. The patients are grouped according to the DNA chip predicted prognosis (Good, Intermediate, Poor) based on the MetaClassifier tool using principal component analysis (PCA).*

negative patients (both clinical markers are normally associated with a Good prognosis) from the four studies where the clinical data is available (Veridex, Stanford, Sotiriou, and Netherlands) shows the ability of the MetaClassifier to further subdivide patients in this category (Figure 11.3).

11.6 SUMMARY

A number of medium- to large-sized studies of breast cancer with DNA microarrays have been published. All are retrospective studies, where few patients have received adjuvant therapy. All but one group have identified genes correlated to outcome, and meta-analysis has shown good agreement between the genes found in some studies, and poor agreement to other studies. Questions have been raised as to the design and representativeness of several studies. Further studies are needed to resolve these differences. But a meta-analysis of the publically available studies show a good agreement. At least 1500 patients have been included in the studies so far.

FURTHER READING

Bertucci, F., Borie, N., Ginestier, C., Groulet, A., Charafe-Jauffret, E., Adelaide, J., Geneix, J., Bachelart, L., Finetti, P., Koki, A., Hermitte, F., Hassoun, J., Debono, S., Viens, P., Fert, V., Jacquemier, J., and Birnbaum, D. (2004). Identification and validation of an ERBB2 gene expression signature in breast cancers. *Oncogene* 23(14):2564–2575.

Gruvberger, S. K., Ringner, M., Eden, P., Borg, A., Ferno, M., Peterson, C., and Meltzer, P. S. (2002). Expression profiling to predict outcome in breast cancer: the influence of sample selection. *Breast Cancer Res.* 5(1):23–26.

Gruvberger-Saal, S. K., Eden, P., Ringner, M., Baldetorp, B., Chebil, G., Borg, A., Ferno, M., Peterson, C., and Meltzer, P. S. (2004). Predicting continuous values of prognostic markers in breast cancer from microarray gene expression profiles. *Mol. Cancer Ther.* 3(2):161–168.

Hedenfalk, I., Duggan, D., Chen, Y., Radmacher, M., Bittner, M., Simon, R., Meltzer, P., Gusterson, B., Esteller, M., Kallioniemi, O. P., Wilfond, B., Borg, A., and Trent, J. (2001). Gene expression profiles in hereditary breast cancer. *N. Engl. J. Med.* 244:539–548.

Martin, K. J., Kritzman, B. M., Price, L. M., Koh, B., Kwan, C. P., Zhang, X., Mackay, A., O'Hare, M. J., Kaelin, C. M., Mutter, G. L., Pardee, A. B., and Sager, R. (2000). Linking gene expression patterns to therapeutic groups in breast cancer. *Cancer Res.* 60(8):2232–2238.

Mecham, B. H., Klus, G. T., Strovel, J., Augustus, M., Byrne, D., Bozso, P., Wetmore, D. Z., Mariani, T. J., Kohane, I. S., and Szallasi, Z. (2004). Sequence-matched probes produce increased cross-platform consistency and more reproducible biological results in microarray-based gene expression measurements. *Nucleic Acids Res.* 32(9):e74.

Miller, L. D., Smeds, J., George, J., Vega, V. B., Vergara, L., Ploner, A., Pawitan, Y., Hall, P., Klaar, S., Liu, E. T., and Bergh, J. (2005). An expression signature for p53 status in human breast cancer predicts mutation status, transcriptional effects, and patient survival. *Proc. Natl. Acad. Sci. USA* 102(38):13550–13555.

Minn, A. J., Gupta, G. P., Siegel, P. M., Bos, P. D., Shu, W., Giri, D. D., Viale, A., Olshen, A. B., Gerald, W. L., and Massague, J. (2005). Genes that mediate breast cancer metastasis to lung. *Nature* 436(7050):518–524.

Modlich, O., Prisack, H. B., Munnes, M., Audretsch, W., and Bojar, H. (2005). Predictors of primary breast cancers responsiveness to preoperative epirubicin/cyclophosphamide-based chemotherapy: translation of microarray data into clinically useful predictive signatures. *J. Transl. Med.* 3:32.

Onda, M., Emi, M., Nagai, H., Nagahata, T., Tsumagari, K., Fujimoto, T., Akiyama, F., Sakamoto, G., Makita, M., Kasumi, F., Miki, Y., Tanaka, T., Tsunoda, T., and Nakamura, Y. (2004). Gene expression patterns as marker for 5-year postoperative prognosis of primary breast cancers. *J. Cancer Res. Clin. Oncol.* 130(9):537–545.

Pollack, J. R., Sorlie, T., Perou, C. M., Rees, C. A., Jeffrey, S. S., Lonning, P. E., Tibshirani, R., Botstein, D., Borresen-Dale, A. L., and Brown, P. O. (2002). Microarray analysis reveals a major direct role of DNA copy number alteration in the transcriptional program of human breast tumors. *Proc. Natl. Acad. Sci. USA* 99(20):12963–12968.

van't Veer, L. J., Dai, H., van de Vijver, M. J., He, Y. D., Hart, A. A., Bernards, R., and Friend, S. H. (2003). Expression profiling predicts outcome in breast cancer. *Breast Cancer Res.* 5(1):57–58.

Wang, Z. C., Lin, M., Wei, L. J., Li, C., Miron, A., Lodeiro, G., Harris, L., Ramaswamy, S., Tanenbaum, D. M., Meyerson, M., Iglehart, J. D., and Richardson, A. (2004). Loss of heterozygosity and its correlation with expression profiles in subclasses of invasive breast cancers. *Cancer Res.* 64(1):64–71.

Weigelt, B., Glas, A. M., Wessels, L. F., Witteveen, A. T., Peterse, J. L., and van't Veer, L. J. (2003). Gene expression profiles of primary breast tumors maintained in distant metastases. *Proc. Natl. Acad. Sci. USA* 100(26):15901–15905.

Weigelt, B., Wessels, L. F., Bosma, A. J., Glas, A. M., Nuyten, D. S., He, Y. D., Dai, H., Peterse, J. L., and Van't Veer, L. (2005). No common denominator for breast cancer lymph node metastasis. *Br. J. Cancer* 93(8):924–932.

12

Leukemia

Leukemia affects blood-forming cells in the bone marrow. It usually affects cells that end up as white blood cells (leukocytes). The bone marrow contains hematopoietic (blood-forming) stem cells that divide and differentiate into red blood cells, platelets, and white blood cells (Figure 12.1). The white blood cells are divided into granular leukocytes (granulocytes), monocytes, and lymphocytes.

Leukemias that start in granulocytes or monocytes are called myeloid leukemias (or myelocytic or myelogenous leukemia). Leukemias that start in the lymphocytes are called lymphocytic leukemias. Lymphomas also originate in the lymphocytes but, unlike leukemias, they do not develop in the bone marrow; they develop in the lymph nodes or other organs.

Leukemias are further divided into acute and chronic. In acute leukemias, the cells cannot mature properly. Immature leukemia cells continue to reproduce and build up; without treatment, most patients would live only a few months.

In chronic leukemias, the cells mature partly but do not obtain their full function.

So, dividing the leukemias according to the cell of origin and maturation stage, we have the four major subtypes of leukemia:

ALL   Acute lymphocytic leukemia
AML   Acute myeloid leukemia
CLL   Chronic lymphocytic leukemia
CML   Chronic myeloid leukemia

The acute leukemias frequently affect children. They are the childhood leukemias. The chronic leukemias most frequently affect adults. They are the adult leukemias.

The acute lymphocytic leukemias are further subdivided according to the lineage: B-cell and T-cell lineage. Each of these is further subdivided according to the maturity of the leukemic cell within the lineage.

Cancer Diagnostics with DNA Microarrays, By Steen Knudsen
Copyright © 2006 John Wiley & Sons, Inc.

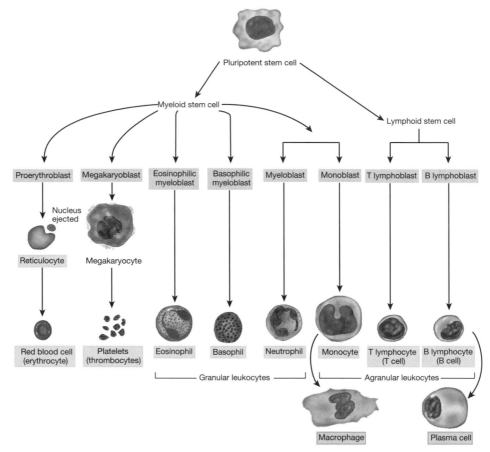

Figure 12.1   *Origin and development of blood cells. (From Tortora, Principles of Human Anatomy, 10th ed., 2005, p. 410. Used with permission of John Wiley & Sons, Inc.)*

12.1   CURRENT DIAGNOSIS

The first examination is blood cell count and blood cell examination, as leukemias affect these. But more detailed information is obtained from a sample of the bone marrow where leukemia originates. The bone marrow can be sampled as a liquid aspirate through a fine needle, or it can be sampled as a larger biopsy through a larger needle.

Cells in the blood and bone marrow are stained and viewed under a microscope by a pathologist. Immunocytochemistry based on antibodies specific to certain blood cells aids in the diagnosis. Cytogenetics is used to look for large-scale chromosomal rearrangements such as translocations and deletions that are found in many leukemias.

12.2   CURRENT THERAPY

Patients are divided into risk groups based on a number of parameters obtained in the diagnosis. The therapy is tailored to the risk group. Chemotherapy is always used, but the dose, time, and type of drug vary. Very high doses of chemotherapy destroy the blood-producing cells in the bone marrow, necessitating a bone marrow or peripheral stem cell transplantation. The stem cells are supplied after the chemotherapy in order to restore blood-forming potential. The stem cells can come either from the patient (removed before chemotherapy and stored) or from another person who does not have leukemia. Stem cell therapy may be needed for patients whose leukemia relapses early after going into remission. It is very expensive (more than $100,000 U.S.).

12.3 MICROARRAY STUDIES OF LEUKEMIA

12.3.1 The Boston Group

A very early paper on the diagnosis of leukemia with microarrays, and in fact one of the papers that showed the potential of microarrays for the diagnosis of cancer, was published by Golub et al. (1999) in *Science*. They used 38 bone marrow samples from patients with acute leukemia (27 ALL and 11 AML) and studied their gene expression with Affymetrix HU6800 GeneChips. Of the 6817 genes on the chip, around 1100 genes were more highly correlated with the ALL/AML class distinction than expected by chance alone.

A class predictor for the ALL and AML classes was built based on the 50 most correlated genes. Class prediction was performed by weighted voting based on the these genes, where the weight is determined by the correlation of each gene. The accuracy obtained was 95% in a cross-validation on the training set of 38 samples and 85% on an independent test set of 34 leukemia samples (24 bone marrow and 10 peripheral blood samples).

In 2002 the group published a study of the oncogenic pathways in T-cell ALL (Ferrando et al., 2002). These findings have clinical relevance, as the group showed that some pathways were associated with a favorable prognosis and others were associated with a poor response to treatment.

In 2002 the group published a study in *Nature Genetics* (Armstrong et al., 2002) of 15 ALL patients harboring translocations involving the *MLL* gene. Using Affymetrix U95A or U95Av2 GeneChips, they showed that the group of patients, displaying a particularly poor prognosis, had an expression pattern that is distinct from other types of ALL. Hence they propose that this constitutes a distinct disease and denote it MLL.

12.3.2 The Austria Group

Stratowa et al. (2001) followed with a study of chronic leukemia. Peripheral blood samples from 54 patients with B-cell chronic lymphocytic leukemia (B-CLL) were analyzed with spotted arrays containing 1237 clones of genes relevant to cancer. They identified a number of genes that showed significant correlation to patient survival and disease staging. These genes were not tested on an independent test set.

12.3.3 The Utah Group

Moos et al. (2002) published a study of 51 patients with childhood leukemia using a 4608-gene cDNA array. Eight samples were from peripheral blood and 43 samples were from bone marrow. Some of them were paired, and clustering revealed that the peripheral blood sample and the bone marrow sample in each case clustered together, indicating the homogeneity of blasts in peripheral blood and bone marrow. Using t-tests they identified genes that separated ALL from AML, B-lineage ALL from T-lineage ALL, B-lineage standard risk versus B-lineage high risk, and *TEl-AML1* fusion versus patients without a translocation. The selected genes were assessed for their classification performance using LOOCV and KNN classifiers. The ALL versus AML classification was 87% correct. T- versus B-lineage classification was 90% correct.

12.3.4 The Memphis Group

Yeoh et al. (2002) published a study of 360 pediatric ALL patients using the Affymetrix U95Av2 GeneChip. Cryopreserved mononuclear cell suspensions from bone marrow aspirates or peripheral blood were used in this study. Clustering of the samples based on their expression profile resulted in six subgroups that coincided with subgroups already identified by immunochemistry and genetic analysis. A seventh subgroup not identified by conventional methods was also observed.

A classification of leukemias into subtypes was attempted by selecting genes using correlation-based feature selection, making decisions using support vector machines and implementing the framework in a decision tree. Unfortunately, the feature selection was performed on the "blinded test set," making the performance estimates unreliable. Within individual subtypes, Yeoh and co-workers were able to identify gene expression profiles that correlated with outcome (relapse). The prediction accuracy of 97–100 % obtained in some subtypes, however, was again overestimated by feature selection on the test set. Later, 132 of the 360 samples were reanalyzed on the new Affymetrix U133 GeneChip platform (Ross et al. 2003).

The group later published a study (Ross et al., 2004) of 130 pediatric and 20 adult cases of AML using the Affymetrix U133A GeneChip. Class discriminating genes were obtained for each of the major prognostic subgroups of AML. A classifier was built and it obtained classification accuracies above 93%. A prognostic score function based on the expression of two probe sets related to relapse was tested as a predictor of outcome and found to be statistically significant.

12.3.5 The Japan Group

Yagi et al. (2003) published a study of 54 pediatric AML patients using Affymetrix U95Av2 GeneChips. They used the t-test to identify genes associated with prognosis. Thirty-five genes separated a test set into two groups with differing prognoses. This gene set was tested by the Munich group (see Section 12.3.6) and not found to differentiate patients according to prognosis in the Munich set.

12.3.6 The Munich Group

Kohlmann et al. (2003) published a study of 90 adult acute leukemias (25 ALL and 65 AML) studied with Affymetrix U95A and U133A GeneChips. Using a similar approach as that used by the Boston group, they identified a set of genes that could discriminate between eight clinically relevant subtypes. They later showed that microarray findings from pediatric ALL can be transferred to adult ALL (using 34 U133A chips; Kohlmann et al., 2004). The group has also showed that 37 adult AML patients can be separated into groups largely corresponding to known chromosomal lesions (using U95A chips; Schoch et al., 2002).

12.3.7 The Stanford Group

Bullinger et al. (2004) published a study of 116 adults with AML using cDNA arrays. Unsupervised clustering revealed new prognostically relevant subgroups of AML. An overall clinical-outcome predictor was built based on 133 genes. In an independent test set, the outcome predictor was a strong independent prognostic factor (multivariate analysis). Thus it improves the molecular classification of adult AML.

12.3.8 The Copenhagen Group

Willenbrock et al. (2004) published a study of 45 children with ALL analyzed with Affymetrix Focus GeneChips. *t*-Tests were used to select genes that could differentiate between pre-B- and T-lineage ALL and between five-year survival and relapse. Six different classification methods were applied and compared, and the best performance achieved was 100% for lineage and 78% correct prediction of event-free survival (Matthews correlation coefficient 0.59). The lineage classifier was tested on a previously published study of 132 samples tested on the Affymetrix U133A GeneChip (Ross et al. 2003). It correctly classified 132 samples.

12.3.9 The Netherlands Group

Valk et al. (2004) published a study of 285 adults with AML using the Affymetric U133A GeneChip. Unsupervised clustering revealed 16 groups, some of which coincided with known chromosomal lesions and mutations. A novel cluster was associated with a poor treatment outcome of the disease.

12.4 SUMMARY

A large number of gene expression studies of leukemia have been performed. They all show that it is possible to distinguish between subtypes of cancer, and they have even identified a proposed novel subtype, MLL, displaying a particularly poor prognosis. Prediction of outcome has also been successful in those studies where it has been tried. It has even been shown that, for some subtypes, such a predictor is better than existing prognosticators. Since all patients receive chemotherapy, the predictors actually predict treatment response or chemosensitivity. DNA microarray may be ready to be deployed for the diagnosis of some subtypes of leukemia. At least 1400 patients have been studied so far.

FURTHER READING

Cario, G., Stanulla, M., Fine, B. M., Teuffel, O., Neuhoff, N. V., Schrauder, A., Flohr, T., Schafer, B. W., Bartram, C. R., Welte, K., Schlegelberger, B., and Schrappe, M. (2005). Distinct gene expression profiles determine molecular treatment response in childhood acute lymphoblastic leukemia. *Blood* 105(2):821–826.

Cheok, M. H., Yang, W., Pui, C. H., Downing, J. R., Cheng, C., Naeve, C. W., Relling, M. V., and Evans, W. E. (2003). Treatment-specific changes in gene expression discriminate in vivo drug response in human leukemia cells. *Nature Genet.* 34(1):85–90. Erratum in: *Nature Genet.* 2003 Jun;34(2):231.

Davis, R. E., and Staudt, L. M. (2002). Molecular diagnosis of lymphoid malignancies by gene expression profiling. *Curr. Opin. Hematol.* 9(4):333–338 (review).

Durig, J., Nuckel, H., Huttmann, A., Kruse, E., Holter, T., Halfmeyer, K., Fuhrer, A., Rudolph, R., Kalhori, N., Nusch, A., Deaglio, S., Malavasi, F., Moroy, T., Klein-Hitpass, L., and Duhrsen, U. (2003). Expression of ribosomal and translation-associated genes is correlated with a favorable clinical course in chronic lymphocytic leukemia. *Blood* 101(7):2748–2755.

Ferrando, A. A., and Thomas Look, A. (2003). Gene expression profiling: will it complement or replace immunophenotyping? *Best Pract. Res. Clin. Haematol.* 16(4):645–652.

Fine, B. M., Stanulla, M., Schrappe, M., Ho, M., Viehmann, S., Harbott, J., and Boxer, L. M. (2004). Gene expression patterns associated with recurrent chromosomal translocations in acute lymphoblastic leukemia. *Blood* 103(3):1043–1049. Epub 2003 Oct 02.

Greiner, T. C. (2004). mRNA microarray analysis in lymphoma and leukemia. *Cancer Treat. Res.* 121:1–12 (review).

Gutierrez, N. C., Lopez-Perez, R., Hernandez, J. M., Isidro, I., Gonzalez, B., Delgado, M., Ferminan, E., Garcia, J. L., Vazquez, L., Gonzalez, M., and San Miguel, J. F. (2005). Gene expression profile reveals deregulation of genes with relevant functions in the different subclasses of acute myeloid leukemia. *Leukemia* 19(3):402–409.

Hayashi, Y. (2003). Gene expression profiling in childhood acute leukemia: progress and perspectives. *Int. J. Hematol.* 78(5):414–420 (review).

Haferlach, T., Kohlmann, A., Kern, W., Hiddemann, W., Schnittger, S., and Schoch, C. (2003). Gene expression profiling as a tool for the diagnosis of acute leukemias. *Semin. Hematol.* 40(4):281–295 (review).

Holleman, A., Cheok, M. H., den Boer M. L., Yang, W., Veerman, A. J., Kazemier, K. M., Pei, D., Cheng, C., Pui, C. H., Relling, M. V., Janka-Schaub, G. E., Pieters, R., and Evans, W. E. (2004). Gene-expression patterns in drug-resistant acute lymphoblastic leukemia cells and response to treatment. *N. Engl. J. Med.* 351(6):533–542.

Kaneta, Y., Kagami, Y., Katagiri, T., Tsunoda, T., Jin-nai, I., Taguchi, H., Hirai, H., Ohnishi, K., Ueda, T., Emi, N., Tomida, A., Tsuruo, T., Nakamura, Y., and Ohno, R. (2002). Prediction of sensitivity to STI571 among chronic myeloid leukemia patients by genome-wide cDNA microarray analysis. *Jpn. J. Cancer Res.* 93(8):849–856.

Lacayo, N. J., Meshinchi, S., Kinnunen, P., Yu, R., Wang, Y., Stuber, C. M., Douglas, L., Wahab, R., Becton, D. L., Weinstein, H., Chang, M. N., Willman, C. L., Radich, J. P., Tibshirani, R., Ravindranath, Y., Sikic, B., and Dahl, G. V. (2004). Gene expression profiles at diagnosis in de novo childhood AML patients identify FLT3 mutations with good clinical outcomes. *Blood* 104(9):2646–2454.

Oshima, Y., Ueda, M., Yamashita, Y., Choi, Y. L., Ota, J., Ueno, S., Ohki, R., Koinuma, K., Wada, T., Ozawa, K., Fujimura, A., and Mano, H. (2003). DNA microarray analysis of hematopoietic stem cell-like fractions from individuals with the M2 subtype of acute myeloid leukemia. *Leukemia* 17(10):1990–1997.

Staal, F. J., van der Burg, M., Wessels, L. F., Barendregt, B. H., Baert, M. R., van den Burg, C. M., van Huffel, C., Langerak, A. W., van der Velden, V. H., Reinders, M. J., and van Dongen, J. J. (2003). DNA microarrays for comparison of gene expression profiles between diagnosis and relapse in precursor-B acute lymphoblastic leukemia: choice of technique and purification influence the identification of potential diagnostic markers. *Leukemia* 17(7):1324–1332. Erratum in: *Leukemia* 2004 May;**18**(5):1041.

Stegmaier, K., Corsello, S. M., Ross, K. N., Wong, J. S., Deangelo, D. J., and Golub, T. R. (2005). Gefitinib (Iressa) induces myeloid differentiation of acute myeloid leukemia. *Blood* 106(8):2841–2848.

Teuffel, O., Dettling, M., Cario, G., Stanulla, M., Schrappe, M., Buhlmann, P., Niggli, F. K., and Schafer, B. W. (2004). Gene expression profiles and risk stratification in childhood acute lymphoblastic leukemia. *Haematologica* 89(7):801–808.

van Delft, F. W., Bellotti, T., Luo, Z., Jones, L. K., Patel, N., Yiannikouris, O., Hill, A. S., Hubank, M., Kempski, H., Fletcher, D., Chaplin, T., Foot, N., Young, B. D., Hann, I. M., Gammerman, A., and Saha, V. (2005). Prospective gene expression analysis accurately subtypes acute leukaemia in children and establishes a commonality between hyperdiploidy and t(12;21) in acute lymphoblastic leukaemia. *Br. J. Haematol.* 130(1):26–35.

13

Lymphoma

Lymphomas originate in lymphatic cells of the lymphoid system. The main types of cells are T lymphocytes and B lymphocytes (see Figure 12.1). The lymphomas are divided into Hodgkin lymphoma and non-Hodgkin lymphoma. This division is for historical reasons; Hodgkin lymphoma is caused by an abnormal B lymphocyte that is easily recognized under a microscope. These cells are called Reed–Sternberg cells. There are several subtypes of Hodgkin lymphoma, but they are all malignant.

All other lymphomas are referred to as non-Hodgkin lymphomas. A large number of different non-Hodgkin lymphomas exist. The most common types are diffuse large B-cell lymphoma (DLBCL), which constitutes about 31% of all lymphomas, and follicular lymphoma (FL), which constitutes about 22% of all lymphomas. The two related diseases, chronic lymphocytic leukemia (CLL) and small lymphocytic lymphoma (SLL), together account for 7% of all lymphomas.

DLBCL is curable in less than 50% of patients.

Lymphoma is diagnosed by histopathological examination of the cells from a biopsy. Immunohistochemistry may be required to distinguish between the individual types of non-Hodgkin lymphoma. Non-Hodgkin lymphomas are staged according to Ann Arbor Staging Systems that divides the lymphomas into four stages according to how much they have spread in the body.

The Ann Arbor Staging System describes the spreading of the disease in stages I–IV. An International Prognostic Index has been developed that takes into account clinical observations—age, stage, spreading, performance status, and serum lactate dehydrogenase levels. It adds one point for each of the five poor prognostic factors: 0–1 means low, 2–3 means medium, and 4–5 means high risk.

The treatment of non-Hodgkin lymphoma follows the standard cancer therapies like radiation therapy, chemotherapy, immunotherapy, and bone marrow transplantation.

Cancer Diagnostics with DNA Microarrays, By Steen Knudsen
Copyright © 2006 John Wiley & Sons, Inc.

13.1 MICROARRAY STUDIES OF LYMPHOMA

13.1.1 The Stanford Group

Alizadeh et al. (2000) published a *Nature* paper on microarray analysis of the three most common types of lymphoma. A special "Lymphochip" was designed, containing 12,069 genes from germinal B-cell library, additional genes from libraries created from specific lymphomas, and genes known to be involved in cancer. In total, 17,856 cDNA clones were included on the Lymphochip. This chip was applied to 96 normal and malignant samples including the three most common types of non-Hodgkin lymphomas: diffuse large B-cell lymphoma (DLBCL), follicular lymphoma (FL), and chronic lymphocytic leukemia (CLL). Hierarchical clustering revealed two distinct subtypes of DLBCL, hitherto unrecognized in clinical practice. One subtype was referred to as germinal center B-like DLBCL, the other activated B-like DLBCL. Patients with the former subtype had a significantly better overall survival than patients with the latter subtype.

In 2001 the same group published results with a reduced set of six genes measured by quantitative real-time PCR (Lossos et al., 2001). The performance of the six genes for classification was validated in a test set and was found to be independent of the International Prognostic Index and added to its power.

13.1.2 The Boston Group

Shipp et al. (2002) published a *Nature Medicine* paper detailing the study of 58 DLBCL and 19 FL patients using the Affymetrix HuGeneFL GeneChip. They built a weighted-voting classifier to classify outcome. In cross-validation tests the classifier produced two groups of patients with very different five-year overall survival rates (70% versus 12%). The performance was distinct from the International Prognostic Index. They compared their findings to the Alizadeh results. Hierarchical clustering based on some of the same genes that Alizadeh had used on their Lymphochip resulted in similar division into germinal center B-like and activated B-like clusters. These clusters however, did not have any significant difference in outcome, suggesting that the association between clusters and outcome in the Alizadeh cohort may have been indirect or incidental. Going the other way, however, using the classifier built on the Shipp data to predict outcome in the Alizadeh data was more successful.

13.1.3 The NIH Group

Rosenwald et al. (2003a) published a large study of 240 DLBCL patients studied with the Lymphochip. In addition to the previously identified subgroups of DLBCL, they discovered a third group and called it type 3 DLBCL. Seventeen genes were selected for a classifier of outcome that was shown to be independent of the International Prognostic Index.

In 2003 the group published an update (Wright et al., 2003), where they used a Bayes rule predictor of membership of one of the two major DLBCL subgroups. They applied this predictor to the Boston group dataset and found that, with this subgrouping, the two clusters had significant difference in outcome.

13.1.4 The NCI Group

Dave et al. (2004) published a study of 191 follicular lymphoma patients using Affymetrix HG-U133A arrays. An expression pattern associated with the length of survival was determined in a training set of 95 specimens. A molecular predictor of survival was constructed from these genes and validated in an independent test set of 96 specimens.

The predictor was later criticized by Robert Tibshirani (2005) as being a "fragile result." He was not able to find any association between gene expression and survival in the dataset using standard methods.

13.2 META-CLASSIFICATION OF LYMPHOMA

In the review of breast cancer studies, a meta-classifier based on principal components was shown. Joining datasets via their principal components is a supervised process: the class labels of the samples are used both for gene selection and for adjusting the signs of the components. It is possible, however, to extract and join components in a completely unsupervised manner, independent of the class relationships of the samples. Independent component analysis (ICA) is very well suited for this. The main difference between ICA and PCA is that ICA extracts statistically independent components that are non-Gaussian, whereas components extracted by PCA can be pure noise (Gaussian). One advantage of ICA in this context is that it is possible to skip the gene selection step that is often used before PCA to reduce noise or uninteresting variation. Instead we can just eliminate genes that have no variation at all, either because they are not expressed in any of the samples, or because their expression is constant.

Independent component analysis is computationally much more complex than PCA: it uses iteration from a random starting point to arrive at the final components.

As an example, an ICA-based MetaClassifier was built for predicting how well lymphoma patients respond to chemotherapy.

Three studies have made their raw data available: NIH, Stanford, and Boston. All datasets were normalized with logit. For each study the 500 genes with the maximum variance were extracted and 7 independent components were extracted using the R implementation of fastICA (Hyvärinen, 1999), without row normalization. The performance is not influenced significantly whether 500 or 1000 genes are used or whether 5 or 10 components are extracted.

For each sample, the projections on the 7 components were extracted from the estimated mixing matrix A:

$$X = SA,$$

where X is the transposed expression matrix of genes versus samples, S contains the independent components, and A is a linear mixing matrix. We assume that independent components correspond to fundamental biological processes or pathways, and that each component describes all the genes that participate in one such process or pathway. The advantage is that not all genes from that pathway or process are necessary to measure the activity of the component, making it possible to extract components from different array platforms with different subsets of genes and to compare them afterwards.

13.2.1   Matching of Components

The order and sign of the independent components are arbitrary and have to be matched across the three datasets. We perform this matching based on the above assumption that each component describes the genes that participate in some biological process or pathway, or describes genes that are coordinately expressed. Thus, despite the severe limitations in matching genes between platforms, we should be able to identify similar components because they have more genes in common than dissimilar components. The genes for each platform are converted to their RefSeq IDs and matched to the genes on the other platforms based on this. Now we are able to calculate a correlation coefficient between components, because each component is merely a weighted sum of genes. We calculate the Pearson correlation coefficient between the weights for those genes that have been matched across platforms using RefSeq. Those components that have the highest correlation coefficient are matched. The sign of one component is adjusted if its highest correlation coefficient to another component is negative. Components that cannot be matched to components extracted from other platforms with a correlation coefficient higher than 0.3 or below −0.3 are discarded.

It is important to realize that while the many false negatives and false positives obtained when matching genes across platforms make it difficult to build a gene-based classifier, they do not prevent us from matching components, because all we need is a difference in correlation coefficient. This difference can be detected even in the presence of false negatives and false positives.

Figure 13.1 shows two such independent components from the three datasets after matching.

Four different classification methods were trained on the matched independent components: K Nearest Neighbor, Nearest Centroid, Support Vector Machine, and Neural Network. The classification was decided by voting among the four methods. Unanimous Good prognosis classification resulted in a Good prognosis prediction. Unanimous Poor prognosis classification resulted in a Poor prognosis prediction. Whenever there was disagreement between the methods, the Intermediate prognosis was predicted.

Testing of the performance of the classifier was done using leave-one-out cross-validation. One at a time, one patient (test sample) from one platform was left out of the independent component matching as well as training of the four classifiers. Then the 7 independent components from the test sample were input to four classifiers and used to predict the prognosis of the test sample. This entire procedure was repeated for all samples until a prediction had been obtained for all. The resulting prediction was plotted according to the clinical outcome (death or survival including censorship) in a Kaplan–Meier plot (Figure 13.2).

When the outcome is shown for patients with IPI risk index low (0–1) in Figure 13.3, IPI index medium (2–3) in Figure 13.4, and IPI index high (4–5) in Figure 13.5, it is seen that the chip prediction is independent of IPI risk index and indeed allows a reclassification of patients (although the log-rank test for IPI low shows that the reclassification is barely significant). Thus, some patients with IPI low risk can be assigned a Poor prognosis and some patients with IPI medium risk can be assigned a Good prognosis. This reclassification can be very important for clinical treatment choices.

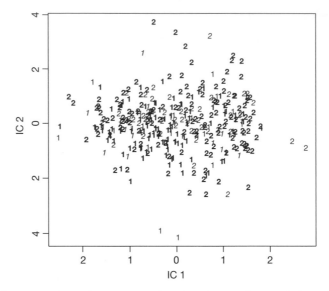

Figure 13.1 *All samples from the three studies plotted according to their projections on two independent components after these have been matched across datasets. Plain font: Boston dataset; bold: NIH dataset; italics: Stanford dataset. The number "1" corresponds to those who survived the disease; the number "2" corresponds to those who died from the disease. A slight deviation from random is seen for the distribution of the two classes. There is a slight overrepresentation of class "1" in the lower left quadrant (both IC1 and IC2 negative). This subtle signal, together with the subtle signals present in the eight remaining components, is used by the classifier to predict outcome (see color insert).*

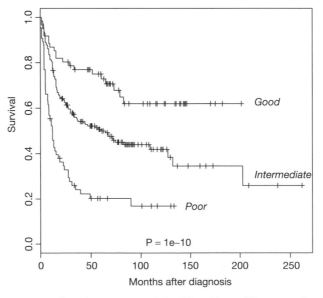

Figure 13.2 *Survival of 334 lymphoma patients joined from three different studies. The patients are grouped according to the DNA chip predicted prognosis (Good, Intermediate, Poor) based on the MetaClassifier tool using independent component analysis (ICA) instead of PCA.*

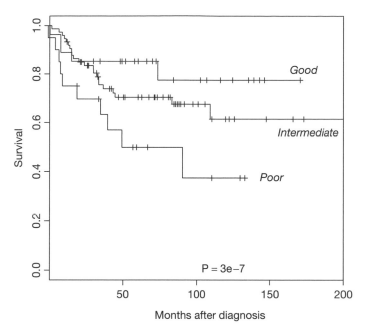

Figure 13.3 *Survival of IPI low risk lymphoma patients predicted by DNA chips. The patients are grouped according to the DNA chip predicted prognosis (Good, Intermediate, Poor) based on the MetaClassifier tool using independent component analysis (ICA) instead of PCA.*

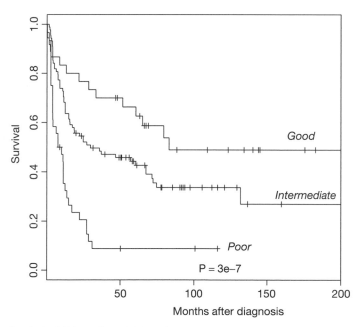

Figure 13.4 *Survival of IPI medium risk lymphoma patients predicted by DNA chips. The patients are grouped according to the DNA chip predicted prognosis (Good, Intermediate, Poor) based on the MetaClassifier tool using independent component analysis (ICA) instead of PCA.*

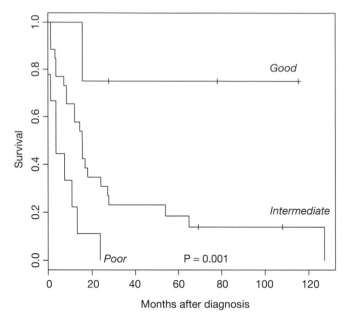

Figure 13.5 *Survival of IPI high risk lymphoma patients predicted by DNA chips. The patients are grouped according to the DNA chip predicted prognosis (Good, Intermediate, Poor) based on the MetaClassifier tool using independent component analysis (ICA) instead of PCA.*

13.2.2   Blind Test Set

The leave-one-out cross-validation shown above uses information from all platforms to classify the left out sample. That means that the remaining samples from the platform where the left out sample comes from are used for predicting the prognosis. So this method does not fully test how well the extracted independent components can be used to classify across platforms. If we instead put two platform datasets in the training set and use them to predict all the samples from the third platform dataset (blind test set), we directly demonstrate the ability of components to be used for classification across platforms.

We still need to perform ICA on the test set and to determine the sign and order of the components between the training set and test set. But this is done without the use of the class labels, and thus the test can be performed on a blind test set.

Figure 13.6 shows the results of predicting all the samples from the Boston datasets using a classifier built on the Stanford and NIH datasets. For comparison, Figure 13.7 shows the result of predicting the Boston samples using a classifier built on all three datasets. The performance is almost the same, underscoring the ability of independent components to transfer information necessary for classification across platforms.

Even though the components are extracted in an unsupervised manner, the information they extract is disease or cell type specific. Components extracted from lung tissue or blood cells in asthmatic patients cannot be used to predict the prognosis of lymphoma patients (data not shown).

One word of caution is important. The fastICA algorithm is nondeterministic. It is not guaranteed to find the same components every time it is run. It can get stuck in local minima. For that reason it is important to run the algorithm several times to ensure

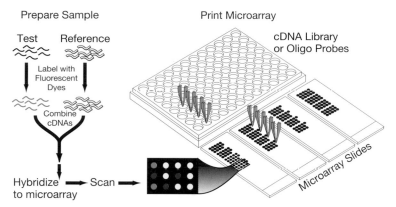

Prepare Sample

Test   Reference

Label with
Fluorescent
Dyes

Combine
cDNAs

Hybridize → Scan →
to microarray

Print Microarray

cDNA Library
or Oligo Probes

Microarray Slides

Figure 1.4 *The spotted array technology. A robot is used to transfer probes in solution from a microtiter plate to a glass slide where they are dried. Extracted mRNA from cells is converted to cDNA and labeled fluorescently. Reference sample is labeled red and test sample is labeled green. After mixing, they are hybridized to the probes on the glass slide. After washing away unhybridized material, the chip is scanned with a confocal laser and the image is analyzed by computer.*

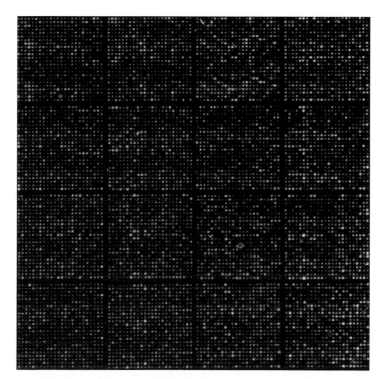

Figure 1.5 *Spotted array containing more than 9000 features. Probes against each predicted open reading frame in Bacillus subtilis are spotted twice on the slide. Image shows color overlay after hybridization of sample and control and scanning. (Picture by Hanne Jarmer.)*

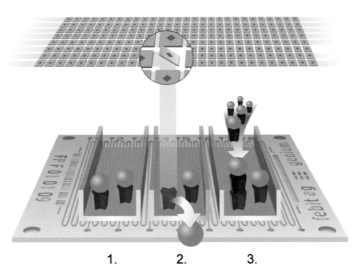

Figure 1.7 *Graphical illustration of the in situ synthesis of probes inside the Febit DNA processor. Shown are three enlargements of a microchannel, each illustrating one step in the synthesis: 1 — the situation before synthesis; 2 — selected positions are deprotected by controlling light illumination via a micromirror; 3 — substrate is added to the microchannel and covalently attached to the deprotected positions. (Copyright Febit AG. Used with permission.)*

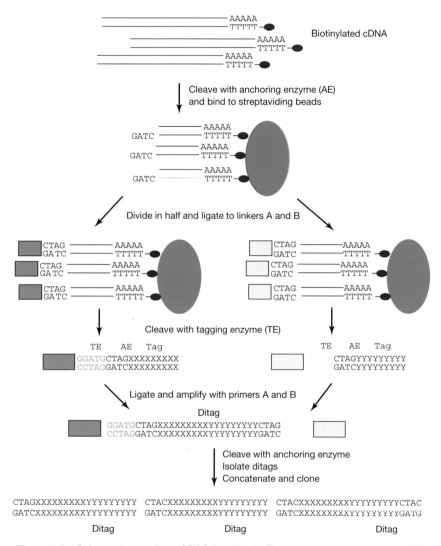

Figure 1.8 *Schematic overview of SAGE methods. (Based on Velculescu et al., 1995.)*

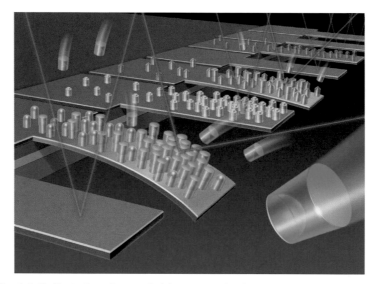

Figure 1.10 *Artist's illustration of array of eight nanomechanical cantilevers. Binding of targets leads to bending that is detected by deflection of a laser beam. (From Concentris. Used with permission.)*

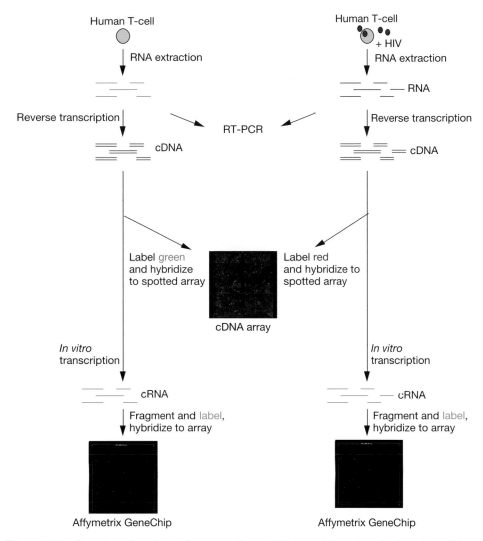

Figure 1.11 Overview of methods for comparing mRNA populations in cells from two different conditions.

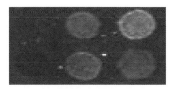

Figure 2.3 Illustration of "ghost" where the background has higher intensity than the spot. In this case, subtracting background from spot intensity may be a mistake.

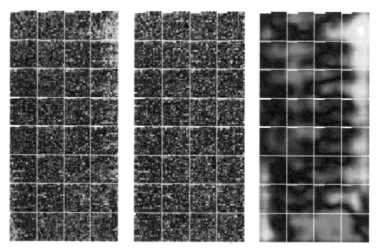

Figure 2.4 *Spatial effects on a spotted array. The blue-yellow color scale indicates fold change between the two channels. A spatial bias is visible (left). Gaussian smoothing captures the bias (right), which can then be removed by subtraction from the image (center). (From Workman et al., 2002).*

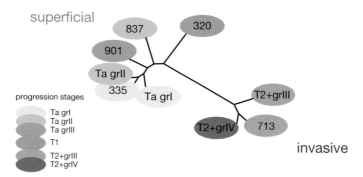

Figure 5.2 *Hierarchical clustering of bladder cancer patients using an unrooted tree. The clustering was based on expression measurements from a DNA array hybridized with mRNA extracted from a biopsy. Numbers refer to patients and the severity of the disease is indicated by a grayscale code. (Christopher Workman, based on data published in Thykjaer et al., 2001.)*

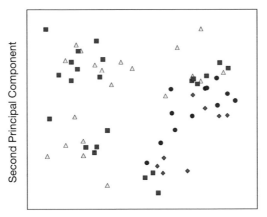

Figure 6.3 *Principal component analysis of 63 small, round blue cell tumors. Different symbols are used for each of the four categories as determined by classical diagnostics tests.*

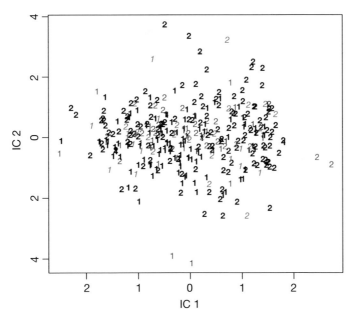

Figure 13.1 *All samples from the three studies plotted according to their projections on two independent components after these have been matched across datasets. Plain font (red): Boston dataset; bold (black): NIH dataset; italics (blue): Stanford dataset. The number "1" corresponds to those who survived the disease; the number "2" corresponds to those who died from the disease. A slight deviation from random is seen for the distribution of the two classes. There is a slight overrepresentation of class "1" in the lower left quadrant (both IC1 and IC2 negative). This subtle signal, together with the subtle signals present in the eight remaining components, is used by the classifier to predict outcome.*

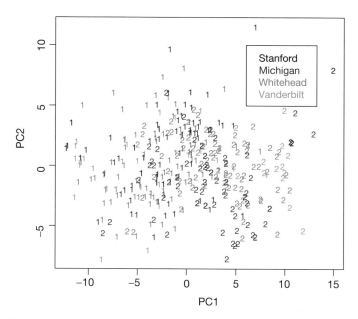

Figure 14.2 *All samples from the four studies are plotted according to their projections on the first two principal components after these have been matched in signs across datasets. Plain font: Stanford dataset; bold: Michigan dataset; italics: Whitehead dataset; bold italics: Vanderbilt dataset. The number "1" corresponds to those who survived the disease; the number "2" corresponds to those who died from the disease. A clear deviation from random is seen for the distribution of the two classes. Patients of class "1" tend to group on one side and patients of class "2" tend to group on the other side. This subtle signal, together with the subtle signals present in the third component, is used by the classifier to predict outcome.*

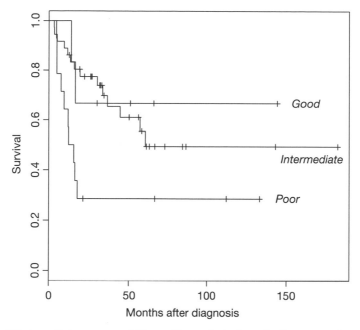

Figure 13.6 *Blind prediction across platforms. The patients from the Boston study were classified blindly using independent components extracted from the Stanford and NIH studies exclusively.*

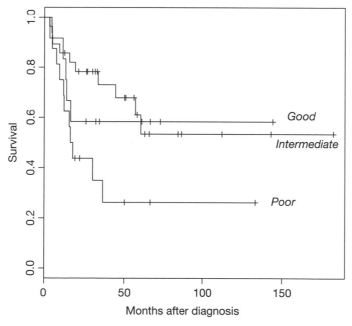

Figure 13.7 *Prediction using all platforms. The patients from the Boston study were classified using independent components extracted from the Stanford, NIH, and Boston studies together.*

that the results obtained from the first run are not uncommonly good or uncommonly poor. They should be reproducible.

13.3 SUMMARY

The application of DNA microarrays to lymphoma has been highly successful. New subgroups have been identified that have a significantly different overall survival. Classifiers to predict survival have been demonstrated to be distinct from currently used prognostic factors and improve the prediction accuracy. All the studies included here have been retrospective studies, and all patients have been treated with anthracycline-based multiagent chemotherapy, CHOP: cyclophosphamide, adriamycine (doxorubicine), vincristine, and prednisone. Thus the prediction of survival is in reality a prediction of response to treatment, or chemosensitivity. At least 415 patients have been included in the studies so far. DNA microarrays appear ready to be deployed for the diagnosis of lymphoma.

FURTHER READING

Dave, S. S., Wright, G., Tan, B., Rosenwald, A., Gascoyne, R. D., Chan, W. C., Fisher, R. I., Braziel, R. M., Rimsza, L. M., Grogan, T. M., Miller, T. P., LeBlanc, M., Greiner, T. C., Weisenburger, D. D., Lynch, J. C., Vose, J., Armitage, J. O., Smeland, E. B., Kvaloy, S., Holte, H., Delabie, J., Connors, J. M., Lansdorp, P. M., Ouyang, Q., Lister, T. A., Davies, A. J., Norton, A. J., Muller-Hermelink, H. K., Ott, G., Campo, E., Montserrat, E., Wilson, W. H., Jaffe, E. S., Simon, R., Yang, L., Powell, J., Zhao, H., Goldschmidt, N., Chiorazzi, M., and Staudt, L. M. (2004). Prediction of survival in follicular lymphoma based on molecular features of tumor-infiltrating immune cells. *N. Engl. J. Med.* 351(21):2159–2169.

Houldsworth, J., Olshen, A. B., Cattoretti, G., Donnelly, G. B., Teruya-Feldstein, J., Qin, J., Palanisamy, N., Shen, Y., Dyomina, K., Petlakh, M., Pan, Q., Zelenetz, A. D., Dalla-Favera, R., and Chaganti, R. S. (2004). Relationship between REL amplification, REL function, and clinical and biologic features in diffuse large B-cell lymphomas. *Blood* 103(5):1862–1868.

Monti, S., Savage, K. J., Kutok, J. L., Feuerhake, F., Kurtin, P., Mihm, M., Wu, B., Pasqualucci, L., Neuberg, D., Aguiar, R. C., Dal, C. I., Ladd, C., Pinkus, G. S., Salles, G., Harris, N. L., Dalla-Favera, R., Habermann, T. M., Aster, J. C., Golub, T. R., and Shipp, M. A. (2005). Molecular profiling of diffuse large B-cell lymphoma identifies robust subtypes including one characterized by host inflammatory response. *Blood* 105(5):1851–1861.

Rosenwald, A., and Staudt, L. M. (2003). Gene expression profiling of diffuse large B-cell lymphoma. *Leuk. Lymphoma* 44 (Suppl 3):S41–S47.

Rosenwald, A., Wright, G., Leroy, K., Yu, X., Gaulard, P., Gascoyne, R. D., Chan, W. C., Zhao, T., Haioun, C., Greiner, T. C., Weisenburger, D. D., Lynch, J. C., Vose, J., Armitage, J. O., Smeland, E. B., Kvaloy, S., Holte, H., Delabie, J., Campo, E., Montserrat, E., Lopez-Guillermo, A., Ott, G., Muller-Hermelink, H. K., Connors, J. M., Braziel, R., Grogan, T. M., Fisher, R. I., Miller, T. P., LeBlanc, M., Chiorazzi, M., Zhao, H., Yang, L., Powell, J., Wilson, W. H., Jaffe, E. S., Simon, R., Klausner, R. D., and Staudt, L. M.

(2003). Molecular diagnosis of primary mediastinal B cell lymphoma identifies a clinically favorable subgroup of diffuse large B cell lymphoma related to Hodgkin lymphoma. *J. Exp. Med.* 198(6):851–862.

Rosenwald, A., Wright, G., Wiestner, A., Chan, W. C., Connors, J. M., Campo, E., Gascoyne, R. D., Grogan, T. M., Muller-Hermelink, H. K., Smeland, E. B., Chiorazzi, M., Giltnane, J. M., Hurt, E. M., Zhao, H., Averett, L., Henrickson, S., Yang, L., Powell, J., Wilson, W. H., Jaffe, E. S., Simon, R., Klausner, R. D., Montserrat, E., Bosch, F., Greiner, T. C., Weisenburger, D. D., Sanger, W. G., Dave, B. J., Lynch, J. C., Vose, J., Armitage, J. O., Fisher, R. I., Miller, T. P., LeBlanc, M., Ott, G., Kvaloy, S., Holte, H., Delabie, J., and Staudt, L. M. (2003). The proliferation gene expression signature is a quantitative integrator of oncogenic events that predicts survival in mantle cell lymphoma. *Cancer Cell* 3(2):185–197.

Savage, K. J., Monti, S., Kutok, J. L., Cattoretti, G., Neuberg, D., De Leval L., Kurtin, P., Dal Cin P., Ladd, C., Feuerhake, F., Aguiar, R. C., Li, S., Salles, G., Berger, F., Jing, W., Pinkus, G. S., Habermann, T., Dalla-Favera, R., Harris, N. L., Aster, J. C., Golub, T. R., and Shipp, M. A. (2003). The molecular signature of mediastinal large B-cell lymphoma differs from that of other diffuse large B-cell lymphomas and shares features with classical Hodgkin lymphoma. *Blood* 102(12):3871–3879.

Valet, G. K., and Hoeffkes, H. G. (2004). Data pattern analysis for the individualised pretherapeutic identification of high-risk diffuse large B-cell lymphoma (DLBCL) patients by cytomics. *Cytometry* 59A(2):232–236.

## 14

Lung Cancer

Air from the trachea is distributed to the bronchi, which divide into the smaller bronchioles and end in tiny air sacs called alveoli (Figure 14.1). In the alveoli the exchange of gases with the bloodstream takes place. Most lung cancers start in the epithelial cells of the bronchi. Lung cancers can also originate in the trachea, bronchioles, and alveoli.

Lung cancers are subdivided into two major groups, small cell lung cancer (SCLC) and non-small cell lung cancer (NSCLC). The non-small cell lung cancers constitute 80% of all lung cancers and are subdivided into three groups: squamous cell carcinoma, adenocarcinoma, and large-cell undifferentiated carcinoma. In addition to SCLC and NSCLC there are more rare types like carcinoid tumors. The different subtypes have different prognoses.

X-ray and biopsy are typical methods to diagnose lung cancer, and after the diagnosis the cancer can be staged according to how much the tumor has spread. This staging system is very similar to that used for breast cancer, following the American Joint Committee on Cancer TNM system. After the staging of the tumor, lymph node, and metastasis, the information is summarized in stages I to IV (Table 14.1).

Treatment of lung cancer is by surgery, radiation therapy, and chemotherapy, either alone or in combination. Stages I and II NSCLC are typically treated by surgery alone, while stages III and IV NSCLC are typically treated by surgery followed by chemotherapy and radiation therapy. In recent years, chemotherapy has been offered to stage I and II patients as well. (See Table 14.2.)

14.1 MICROARRAY STUDIES OF LUNG CANCER

14.1.1 The Harvard/Whitehead Group

Bhattacharjee et al., (2001) published a *PNAS* paper on the study of 186 lung tumors and 17 normal lung samples using the Affymetrix U95A GeneChip. The tumors

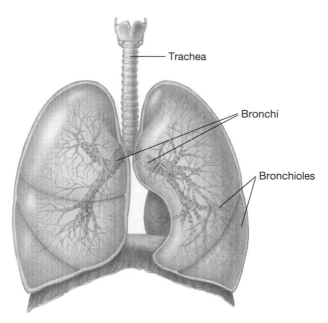

Figure 14.1 *Anatomy of the lung. (From Tortora, Principles of Human Anatomy, 10th ed., 2005, p. 741. Used with permission of John Wiley & Sons, Inc.)*

TABLE 14.1   Summary Staging of Non-Small Cell Lung Cancer

Overall Stage	T Stage	N Stage	M Stage
Stage 0	Tis (in situ)	N0	M0
Stage IA	T1	N0	M0
Stage IB	T2	N0	M0
Stage IIA	T1	N1	M0
Stage IIB	T2	N1	M0
	T3	N0	M0
Stage IIIA	T1	N2	M0
	T2	N2	M0
	T3	N1	M0
	T3	N2	M0
Stage IIIB	Any T	N3	M0
	T4	Any N	M0
Stage IV	Any T	Any N	M1

included 127 adenocarcinomas, 21 squamous cell carcinomas, 20 pulmonary carcinoids, and 6 SCLCs. Twelve adenocarcinomas were suspected to be extrapulmonary metastases. Hierarchical clustering revealed subgroups consistent with the known histological divisions. In addition, two new subgroups of adenocarcinomas were discovered, with differences in prognosis. The study also showed that it is possible to identify lung tumors that are metastases of extrapulmonary origin.

In 2002 a group from the same institutions (Gordon et al., 2002) published a study of 31 malignant pleural mesotheliomas (MPMs) and 150 adenocarcinomas (ADCAs). Thirty-two of the tumor samples were studied using Affymetrix U95A GeneChips.

TABLE 14.2 Non-Small Cell
Lung Cancer Survival by
Stage

Stage	Five-Year Survival
I	47 %
II	26 %
III	8 %
IV	2 %

They identified three pairs of genes, the expression ratio of which was a highly accurate (99%) tool for RT-PCR diagnosis of MPM versus ADCA in a test set of the remaining samples.

14.1.2 The Minnesota Group

A study by Hoang et al. (2004) detailed microarray analysis of 15 non-small cell lung carcinomas. They selected genes able to discriminate between tumors with metastasis and those without. The resulting clusters showed that the genes selected had predictive value on the metastasis of a primary lung carcinoma.

14.1.3 The Vanderbilt Group

Yamagata et al. (2003) published a study of 45 NSCLCs using cDNA microarrays showing that it was possible to select genes that discriminate between primary lung tumors and normal lung and metastasis.

14.1.4 The Tokyo Group

Kikuchi et al. (2003) studied 37 NSCLCs using a cDNA microarray with 23,040 genes. They selected genes that could predict lymph node metastasis of adenocarcinomas. They also selected genes that could predict the sensitivity of NSCLCs to six anticancer drugs.

14.1.5 The Michigan Group

Beer et al. (2002) used univariate Cox analysis to identify genes associated with survival in lung adenocarcinomas. The study was based on 91 lung tumors analyzed with Affymetrix HuGeneFL GeneChip. A risk index based on 50 genes related to survival was validated on an independent sample of lung adenocarcinomas with known clinical outcome. The identification of a set of genes that predict survival in early-stage lung adenocarcinoma allows delineation of a high-risk group that may benefit from adjuvant therapy.

14.1.6 The Mayo Clinic

Sun et al. (2004) studied 15 stage I squamous cell carcinomas using Affymetrix HG-U133A GeneChips. Genes associated with survival were identified using pathway analysis.

14.1.7   The Toronto Group

Wigle et al. (2002) published a study of 39 NSCLCs using a 19,200 cDNA array. The NSCLCs included squamous cell carcinoma, adenocarcinoma, large-cell undifferentiated carcinoma, and carcinoid tumors. Using clustering, they identified clusters with differences in outcome.

14.1.8   The NIH Group

Miura et al. (2002) used a cDNA array with 18,432 clones to investigate 19 patients with adenocarcinoma. The samples were extracted by laser capture microdissection. Forty-five genes were identified that separate smokers from nonsmokers. Twenty-seven genes were identified that separate survivors from nonsurvivors five years after surgery.

14.1.9   The Stanford Group

Garber et al. (2001) studied 67 lung tumors from 56 patients using a 24,000-feature cDNA array. The samples consisted of 41 ACs, 16 SCCs, 5 LCLCs, and 5 SCLCs. Five normal lung specimens were studied. Clustering resulted in subgroups that were in agreement with the known morphological classification. But there were also subgroups that correlated with the survival of the patient.

14.1.10   The Israel Group

Cojocaru et al. (2002) studied 12 NSCLCs and 7 normal lung samples using Affymetrix U95A GeneChips. Clustering revealed a clear separation between the tumors and normal tissue.

14.2   META-CLASSIFICATION OF LUNG CANCER

Four groups have made their raw data available: Whitehead, Vanderbilt, Stanford, and Michigan. The raw intensities from each study were logit normalized and genes were correlated with outcome selected using the t-test. The top 100 ranking genes were subjected to principal component analysis, and the three first principal components for each sample (patient) were retained.

The principal components from all four datasets were combined as shown in Figure 14.2 and used to develop a classifier to predict outcome of the disease of the patient. First the arbitrary signs of the principal components from each platform had to be adjusted. For each principal component we chose the sign that made the sum of squared errors over classes and datasets smallest.

Four different classification methods were trained on the principal components: K Nearest Neighbor (knn algorithm from www.r-project.org), Nearest Centroid, Support Vector Machine (svm algorithm from e1071 package at www.r-project.org), and Neural Network (nnet algorithm with 6 hidden units from nnet package at www.r-project.org). The classification was decided by voting among the four methods: Unanimous Good prognosis classification resulted in a Good prognosis prediction. Unanimous Poor prognosis classification resulted in a Poor prognosis prediction. Whenever there was disagreement between the methods, the Intermediate prognosis was predicted.

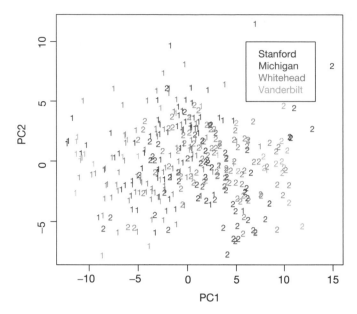

Figure 14.2 *All samples from the four studies are plotted according to their projections on the first two principal components after these have been matched in signs across datasets. Plain font: Stanford dataset; bold: Michigan dataset; italics: Whitehead dataset; bold italics: Vanderbilt dataset. The number "1" corresponds to those who survived the disease; the number "2" corresponds to those who died from the disease. A clear deviation from random is seen for the distribution of the two classes. Patients of class "1" tend to group on one side and patients of class "2" tend to group on the other side. This subtle signal, together with the subtle signals present in the third component, is used by the classifier to predict outcome (see color insert).*

Testing of the performance of the classifier was done using leave-one-out cross-validation. One at a time, one patient (test sample) from one platform was left out of the gene selection and principal component selection as well as training of the five classifiers. Then the genes selected based on the remaining samples were extracted from the test sample and projected onto the principal components calculated based on the remaining samples. The resulting three principal components were input to four classifiers and used to predict the prognosis of the test sample. This entire procedure was repeated for all samples until a prediction had been obtained for all. The resulting prediction was plotted according to the clinical outcome (death or survival including censorship) in a Kaplan–Meier plot (Figure 14.3).

In Figures 14.4, 14.5, and 14.6, the outcomes for the individual stages of lung cancer are shown. It is clear that chip prognosis is independent of clinical staging and can be used to reclassify patients. For example, some stage I patients have a predicted Poor prognosis while some stage II patients have a predicted Good prognosis. This can have an effect on the choice of clinical treatment of the patient.

The leave-one-out cross-validation shown above uses information from all platforms to classify the left out sample. That means that the remaining samples from the platform where the left out sample comes from are used for predicting the prognosis. So this method does not fully test how well the extracted components can be used to classify across platforms. If we instead put three platform datasets in the training set and use

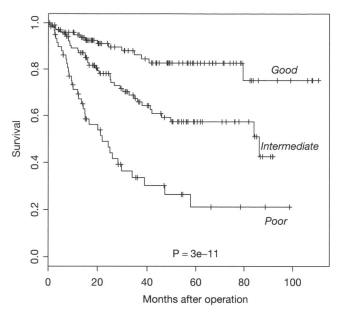

Figure 14.3 *Survival of 269 lung cancer patients combined from four different studies. The patients are grouped according to the DNA chip predicted prognosis (Good, Intermediate, Poor) based on the MetaClassifier tool using principal component analysis (PCA).*

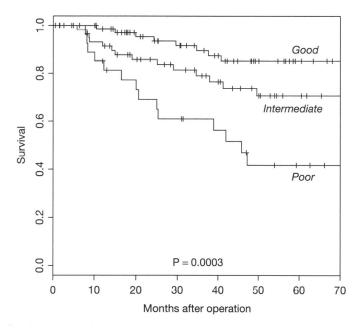

Figure 14.4 *Survival of stage I lung cancer patients combined from four different studies. The patients are grouped according to the DNA chip predicted prognosis (Good, Intermediate, Poor) based on the MetaClassifier tool using principal component analysis (PCA).*

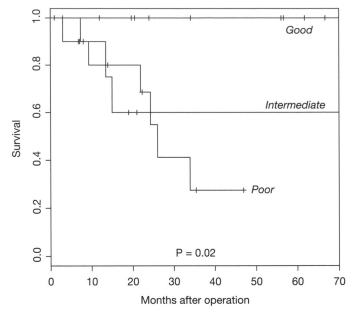

Figure 14.5 *Survival of stage II lung cancer patients combined from four different studies. The patients are grouped according to the DNA chip predicted prognosis (Good, Intermediate, Poor) based on the MetaClassifier tool using principal component analysis (PCA).*

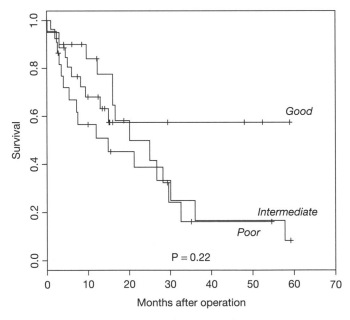

Figure 14.6 *Survival of stage III–IV lung cancer patients combined from four different studies. The patients are grouped according to the DNA chip predicted prognosis (Good, Intermediate, Poor) based on the MetaClassifier tool using principal component analysis (PCA).*

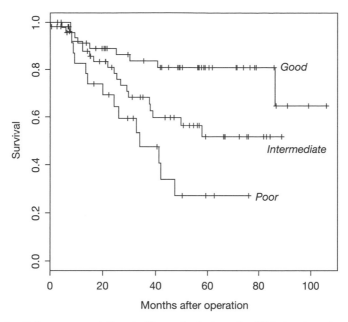

Figure 14.7 *Prediction across platforms. The patients from the Whitehead study were classified using principal components extracted from the Stanford, Vanderbilt, and Michigan studies exclusively.*

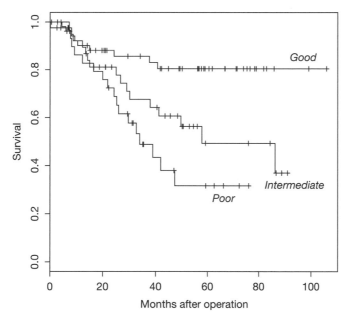

Figure 14.8 *Prediction using all platforms. The patients from the Whitehead study were classified using principal components extracted from the Stanford, Vanderbilt, Michigan, and Whitehead studies together.*

them to predict all the samples from the fourth platform dataset (test set), we directly demonstrate the ability of components to be used for classification across platforms.

We still need to perform a PCA on the test set and to determine the sign of the components between the training set and test set. Because information on the class label of the test set is used for this matching (and also for the t-test selection preceding PCA), it is necessary to loop over the test set with a leave-one-out prediction so the class label of the predicted sample is not used in any way before prediction.

Figure 14.7 shows the results of predicting all the samples from the Whitehead dataset using a classifier built on the Vanderbilt, Stanford, and Michigan datasets. For comparison, Figure 14.8 shows the results of predicting the Whitehead samples using a classifier built on all four datasets. The performance is virtually the same, underscoring the ability of principal components to transfer information necessary for classification across platforms.

14.3   SUMMARY

The application of DNA microarrays to lung cancer has been highly successful. Not only have new subgroups with different prognoses been identified, but several studies have been successful in predicting outcome. At least 575 patients have been included in the studies so far. DNA microarrays appear ready to be used for the diagnosis of lung cancer.

FURTHER READING

Borczuk, A. C., Shah, L., Pearson, G. D., Walter, K. L., Wang, L., Austin, J. H., Friedman, R. A., and Powell, C. A. (2004). Molecular signatures in biopsy specimens of lung cancer. *Am. J. Respir. Crit. Care Med.* 170(2):167–174.

Gordon, G. J., Rockwell, G. N., Godfrey, P. A., Jensen, R. V., Glickman, J. N., Yeap, B. Y., Richards, W. G., Sugarbaker, D. J., and Bueno, R. (2005). Validation of genomics-based prognostic tests in malignant pleural mesothelioma. *Clin. Cancer Res.* 11(12):4406–4414.

Heighway, J., Knapp, T., Boyce, L., Brennand, S., Field, J. K., Betticher, D. C., Ratschiller, D., Gugger, M., Donovan, M., Lasek, A., and Rickert, P. (2002). Expression profiling of primary non-small cell lung cancer for target identification. *Oncogene* 21(50):7749–7763.

Parmigiani, G., Garrett-Mayer, E. S., Anbazhagan, R., and Gabrielson, E. (2004). A cross-study comparison of gene expression studies for the molecular classification of lung cancer. *Clin. Cancer Res.* 10(9):2922–2927.

Talbot, S. G., Estilo, C., Maghami, E., Sarkaria, I. S., Pham, D. K., O-charoenrat, P., Socci, N. D., Ngai, I., Carlson, D., Ghossein, R., Viale, A., Park, B. J., Rusch, V. W., and Singh, B. (2005). Gene expression profiling allows distinction between primary and metastatic squamous cell carcinomas in the lung. *Cancer Res.* 65(8):3063–3071.

Virtanen, C., Ishikawa, Y., Honjoh, D., Kimura, M., Shimane, M., Miyoshi, T., Nomura, H., and Jones, M. H. (2002). Integrated classification of lung tumors and cell lines by expression profiling. *Proc. Natl. Acad. Sci. USA* 99(19):12357–12362.

Yang, P., Sun, Z., Aubry, M. C., Kosari, F., Bamlet, W., Endo, C., Molina, J. R., and Vasmatzis, G. (2004). Study design considerations in clinical outcome research of lung cancer using microarray analysis. *Lung Cancer* 46(2):215–226.

15

Bladder Cancer

The urinary bladder contains a lining of urothelial cells (also called transitional cells). Beneath the urothelium there is a thin layer of connective tissue called the *lamina propria* (Figure 15.1). The next layer is the *muscularis propria*. Finally, fatty connective tissue separates the bladder from other organs.

More than 90% of bladder tumors are urothelial carcinomas. The urothelial carcinomas are further subdivided into the noninvasive tumors and the invasive tumors that have spread to the deeper layers. The superficial, noninvasive tumors sometimes progress to invasive tumors, sometimes they reoccur after removal. So they need to be monitored after treatment.

The diagnosis is made using a cytoscope inserted into the bladder through the urethra. Suspicious growths can be examined by taking a biopsy. Histopathological examination of the biopsy will reveal if there is cancer. Pathological staging assigns the tumor to one of the following categories (American Joint Committee on Cancer):

TX Primary tumor cannot be assessed due to lack of information.

T0 No evidence of primary tumor.

Ta Noninvasive papillary carcinoma.

Tis Carcinoma in situ (CIS); noninvasive flat carcinoma.

T1 Tumor has spread to subepithelial connective tissue.

T2 Tumor has spread to muscle.

T2a Tumor has spread to superficial muscle (inner half).

T2b Tumor has spread to deep muscle (outer half).

T3 Tumor has spread to perivesical tissue (fatty tissue that surrounds the bladder).

Cancer Diagnostics with DNA Microarrays, By Steen Knudsen
Copyright © 2006 John Wiley & Sons, Inc.

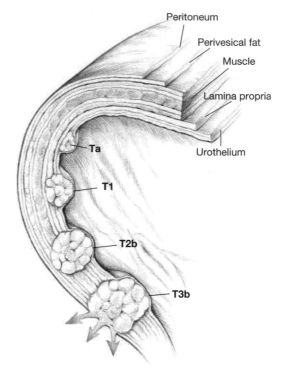

Figure 15.1   *The layers of the bladder and the different stages of bladder cancer. (Illustration by Tim Phelps. Copyright 1999 BPH Guideline, American Urological Association. Used with permission.)*

T3a   Tumor has spread to perivesical tissue microscopically (tumor deposits in perivesical tissue are very small and cannot be seen without a microscope).

T3b   Tumor has spread to perivesical tissue macroscopically (tumor deposits large enough to be recognized by imaging tests or to be seen or felt by the surgeon).

T4   Tumor has spread to any of the following: prostate, uterus, vagina, pelvic wall, or abdominal wall.

T4a   Tumor has spread to the prostate, uterus, and/or vagina.

T4b   Tumor has spread to the pelvic wall or the abdominal wall.

Further staging according to lymph node infiltration and metastasis is available.

Treatment of bladder cancer consists of surgery, radiation therapy, immunotherapy, and chemotherapy. The choice of one or more of these treatments depends on the staging of the cancer. For example, surgery can range from a simple transurethral resection of a superficial tumor to a radical cystectomy removing the entire bladder and nearby organs for large invasive tumors.

15.1   MICROARRAY STUDIES OF BLADDER CANCER

15.1.1   The Aarhus Group

Thykjaer et al. (2001) published a study of 39 bladder cancer biopsies compared to a pool of 36 normal bladder tissue samples. Hierarchical clustering revealed subgroups that coincided with the histopathological staging of the disease.

In 2002 a classifier was published in *Nature Genetics* by Dyrskot et al. (2002). Based on 40 tumor biopsies, 32 genes were selected. A maximum likelihood classifier based on the 32 genes was tested on an independent set of 68 tumors that were analyzed on a cDNA array and was able to classify correctly 84% of the Ta tumors, 50% of the T1 tumors, and 74% of the T2+ tumors. An outcome predictor was developed for superficial Ta tumors. It was able to predict recurrence with 75% accuracy in a set of 31 Ta tumors.

15.1.2 The New York Group

Sanchez-Carbayo et al. (2003) published a study of 9 bladder cancer cell lines analyzed with a 8976-clone cDNA array and 15 solid bladder cancer tumors analyzed with a 17,842-clone cDNA array. The gene *KiSS-1* was found differentially expressed in both studies and its expression was found to be significantly associated with overall survival four years after diagnosis. This finding was confirmed by in situ hybridization and tissue microarrays.

15.1.3 The Dusseldorf Group

Modlich et al. (2004) studied 22 superficial and 20 invasive bladder cancer specimens using filter-based cDNA arrays with 1185 genes (Atlas Human Cancer 1.2). Nine of the specimens were verified using Affymetrix U133A GeneChips. Two-way clustering using different subsets of genes grouped tumor samples according to clinical outcome as superficial, invasive, or metastasizing. Furthermore, the study revealed a clonal origin of superficial tumors and revealed that samples taken from different regions inside a tumor had highly similar gene expression.

15.2 SUMMARY

The number of clinical microarray studies on bladder cancer is limited. There is some suggestion that the outcome (measured either as recurrence of Ta or as overall survival in all stages) can be predicted. Further studies with larger and different patient populations are needed to verify these observations. At least 176 patients have been studied so far.

FURTHER READING

Blaveri, E., Brewer, J. L., Roydasgupta, R., Fridlyand, J., Devries, S., Koppie, T., Pejavar, S., Mehta, K., Carroll, P., Simko, J. P., and Waldman, F. M. (2005). Bladder cancer stage and outcome by array-based comparative genomic hybridization. *Clin. Cancer Res.* 11(19):7012–7022.

Blaveri, E., Simko, J. P., Korkola, J. E., Brewer, J. L., Baehner, F., Mehta, K., Devries, S., Koppie, T., Pejavar, S., Carroll, P., and Waldman, F. M. (2005). Bladder cancer outcome and subtype classification by gene expression. *Clin. Cancer Res.* 11(11):4044–4055.

Duggan, B. J., McKnight, J. J., Williamson, K. E., Loughrey, M., O'Rourke, D., Hamilton, P. W., Johnston, S. R., Schulman, C. C., and Zlotta, A. R. (2003). The need to embrace molecular profiling of tumor cells in prostate and bladder cancer. *Clin. Cancer Res.* 9(4):1240–1247 (review).

Dyrskjot, L. (2003). Classification of bladder cancer by microarray expression profiling: towards a general clinical use of microarrays in cancer diagnostics. *Expert. Rev. Mol. Diagn.* 3(5):635–647 (review).

Dyrskjot, L., Zieger, K., Kruhoffer, M., Thykjaer, T., Jensen, J. L., Primdahl, H., Aziz, N., Marcussen, N., Moller, K., and Orntoft, T. F. (2005). A molecular signature in superficial bladder carcinoma predicts clinical outcome. *Clin. Cancer Res.* 11(11):4029–4036.

Nawrocki, S., Skacel, T., and Brodowicz, T. (2003). From microarrays to new therapeutic approaches in bladder cancer. *Pharmacogenomics* 4(2):179–189 (review).

Nocito, A., Bubendorf, L., Maria, T. I., Suess, K., Wagner, U., Forster, T., Kononen, J., Fijan, A., Bruderer, J., Schmid, U., Ackermann, D., Maurer, R., Alund, G., Knonagel, H., Rist, M., Anabitarte, M., Hering, F., Hardmeier, T., Schoenenberger, A. J., Flury, R., Jager, P., Luc Fehr, J., Schraml, P., Moch, H., Mihatsch, M. J., Gasser, T., and Sauter, G. (2001). Microarrays of bladder cancer tissue are highly representative of proliferation index and histological grade. *J. Pathol.* 194(3):349–357.

Sanchez-Carbayo, M. (2004). Recent advances in bladder cancer diagnostics. *Clin. Biochem.* 37(7):562–571.

Sanchez-Carbayo, M., Socci, N. D., Charytonowicz, E., Lu, M., Prystowsky, M., Childs, G., and Cordon-Cardo, C. (2002). Molecular profiling of bladder cancer using cDNA microarrays: defining histogenesis and biological phenotypes. *Cancer Res.* 62(23):6973–6980.

Sanchez-Carbayo, M., Socci, N. D., Lozano, J. J., Li, W., Charytonowicz, E., Belbin, T. J., Prystowsky, M. B., Ortiz, A. R., Childs, G., and Cordon-Cardo, C. (2003). Gene discovery in bladder cancer progression using cDNA microarrays. *Am. J. Pathol.* 163(2):505–516.

Ying-Hao, S., Qing, Y., Lin-Hui, W., Li, G., Rong, T., Kang, Y., Chuan-Liang, X., Song-Xi, Q., Yao, L., Yi, X., and Yu-Ming, M. (2002). Monitoring gene expression profile changes in bladder transitional cell carcinoma using cDNA microarray. *Urol. Oncol.* 7(5):207–212.

16

Colon Cancer

The small intestine joins the large intestine, also called the colon (Figure 16.1). The final part of the colon is called the rectum which is closed by the anus. More than 95% of colorectal cancers are adenocarcinomas, referring to their origin in the glandular cells that line the inside of the colon and the rectum.

Diagnosis of colorectal cancer is mainly done by colonoscopy, during which a biopsy is removed for histopathological examination under a microscope.

There are three systems for staging colorectal cancer according to how much it has spread through the layers of the colon or rectum and to nearby or distant organs. The American Joint Committee on Cancer (AJCC) System (also called the TNM System) describes stages using roman numerals I through IV. Dukes and Astler-Coller describe stages A through D. The staging is important for treatment, as the prognosis is very different for the different stages:

Stage I   96% five-year relative survival

Stage II   87% five-year relative survival

Stage III   55% five-year relative survival

Stage IV   5% five-year relative survival

Treatment of colorectal cancer consists of surgery, radiation therapy, and chemotherapy.

16.1   MICROARRAY STUDIES OF COLON CANCER

16.1.1   The Princeton Group

Alon et al. (1999) published a study of 40 tumor and 22 normal colon tissue samples using the Affymetrix HU6800 GeneChip. Hierarchical two-way clustering grouped samples into cancerous and noncancerous tissue.

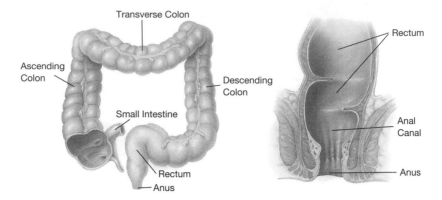

Figure 16.1   *Anatomy of the large intestine. (From Tortora, Principles of Human Anatomy, 10th ed., 2005, p. 795. Used with permission of John Wiley & Sons, Inc.)*

Notterman et al. (2001) published a study of 18 colon adenocarcinomas and 4 adenomas together with paired normal tissue for each of the tumors. Hierarchical clustering again separated the samples into their clinical diagnosis.

16.1.2   The Maryland Group

Zou et al. (2002) published a study of 17 colon tissue samples (9 carcinomas and 8 normal samples) using cDNA microarrays. Unsupervised clustering grouped the samples into carcinomas and normal samples.

16.1.3   The Aarhus Group

Frederiksen et al. (2003) published a study of 20 colorectal tumors (5 samples from each of the Dukes' stages A, B, C, and D) and 5 normal mucosa using Affymetrix HU6800 GeneChips. A nearest neighbor classifier was used to classify normal and Dukes' B and C samples with less than 20% error. Dukes' A and D could not be classified correctly.

16.1.4   The Marseilles Group

Bertucci et al. (2004b) published a study of 50 cancerous and noncancerous colon tissues using cDNA arrays with 8000 clones. Unsupervised hierarchical clustering was able to distinguish normal versus cancer and metastatic versus nonmetastatic tumors. A classifier based on a selected gene set was able to divide patients into two groups with significantly different five-year survival (100% in one group, 40% in the other group).

16.1.5   The San Diego Group

Wang et al. (2004) published a study of 74 patients with Dukes' B colon cancer using Affymetrix U133A GeneChips. A multivariate Cox model was built to predict recurrence within three years. A 23-gene signature of recurrence was tested in an independent set of 36 patients and showed an overall performance accuracy of 78%.

16.1.6   The Helsinki Group

Arango et al. (2005) published a study of Dukes' C colon cancer using Affymetrix HG-U133A GeneChips. A KNN classifier was able to predict recurrence after surgery.

16.2   SUMMARY

Clinical microarray analyses of colon cancers all agree that it is possible to distinguish between colon cancer and tumors of other origin and normal tissue. First efforts have been made at predicting the prognosis of colon cancer, but more studies are necessary. At least 245 patients have been studied so far.

FURTHER READING

Barrier, A., Lemoine, A., Boelle, P. Y., Tse, C., Brault, D., Chiappini, F., Breittschneider, J., Lacaine, F., Houry, S., Huguier, M., Van der Laan, M. J., Speed, T., Debuire, B., Flahault, A., and Dudoit, S. (2005). Colon cancer prognosis prediction by gene expression profiling. *Oncogene* 24(40):6155–6164.

Birkenkamp-Demtroder, K., Olesen, S. H., Sorensen, F. B., Laurberg, S., Laiho, P., Aaltonen, L. A., and Orntoft, T. F. (2005). Differential gene expression in colon cancer of the caecum versus the sigmoid and rectosigmoid. *Gut* 54(3):374–384.

Giordano, T. J., Shedden, K. A., Schwartz, D. R., Kuick, R., Taylor, J. M., Lee, N., Misek, D. E., Greenson, J. K., Kardia, S. L., Beer, D. G., Rennert, G., Cho, K. R., Gruber, S. B., Fearon, E. R., and Hanash, S. (2001). Organ-specific molecular classification of primary lung, colon, and ovarian adenocarcinomas using gene expression profiles. *Am. J. Pathol.* 159(4):1231–1238.

Kruhoffer, M., Jensen, J. L., Laiho, P., Dyrskjot, L., Salovaara, R., Arango, D., Birkenkamp-Demtroder, K., Sorensen, F. B., Christensen, L. L., Buhl, L., Mecklin, J. P., Jarvinen, H., Thykjaer, T., Wikman, F. P., Bech-Knudsen, F., Juhola, M., Nupponen, N. N., Laurberg, S., Andersen, C. L., Aaltonen, L. A., and Orntoft, T. F. (2005). Gene expression signatures for colorectal cancer microsatellite status and HNPCC. *Br. J. Cancer* 92(12):2240–2248.

17

Ovarian Cancer

In women, the ovaries produce eggs for reproduction. There are three main types of ovarian tumors (American Cancer Association):

1. Germ cell tumors start from the cells that produce the ova (eggs) (Figure 17.1).
2. Stromal tumors start from connective tissue cells that hold the ovary together and produce the female hormones estrogen and progesterone.
3. Epithelial tumors start from the cells that cover the outer surface of the ovary.

Epithelial ovarian carcinomas constitute 85–90% of all ovarian cancers and are divided into serous, mucinous, endometrioid, and clear cell types.

More than 50% of women survive more than five years after diagnosis of ovarian cancer.

Diagnosis is by biopsy and various imaging techniques: X-ray, ultrasound, and magnetic resonance. Ovarian cancer, like many other cancers, is staged according to the AJCC/TNM System.

Treatment consists of surgery, radiation therapy, and chemotherapy.

17.1 MICROARRAY STUDIES OF OVARIAN CANCER

17.1.1 Duke University Medical Center

Lancaster et al. (2004) published a study of 42 advanced stage ovarian cancers from patients who survived either less than two years or more than seven years. Gene expression patterns associated with survival were identified.

Cancer Diagnostics with DNA Microarrays, By Steen Knudsen
Copyright © 2006 John Wiley & Sons, Inc.

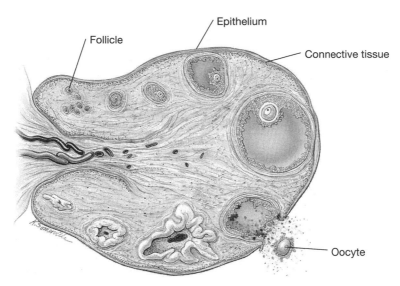

Figure 17.1 *Histology of the ovary. (From Tortora, Principles of Human Anatomy, 10th ed., 2005, p. 853. Used with permission of John Wiley & Sons, Inc.)*

17.1.2 The Stanford Group

Schaner et al. (2003) published a study of 59 archived ovarian cancer samples using three different cDNA microarrays based on IMAGE clones. Thirty-nine samples represented serous papillary carcinomas, 7 clear cell, 2 endometrial, 4 undifferentiated, and 3 adenocarcinomas. A group of genes that distinguished clear cell from other ovarian carcinomas was identified. Clear cell subtype is associated with a worse prognosis and therapeutic resistance. In addition, it was possible to distinguish between breast and ovarian cancer and between different grades of ovarian cancer.

17.1.3 The London Group

Adib et al. (2004) published a study of ovarian cancers using Affymetrix U95A arrays. They found that the expression profile of primary and secondary tumors from the same individual were essentially alike. They also found a biomarker, mammaglobin-2 (MGB2), which is specific to ovarian cancer.

17.1.4 The UCLA Group

Matei et al. (2002) published a study of 9 normal ovaries and 21 epithelial ovarian carcinomas. Hierarchical clustering revealed a clear distinction between normal tissue and carcinoma.

17.1.5 The NIH Group

Sawiris et al. (2002) published a description of their Ovachip, containing 516 cDNAs chosen from serial analysis of gene expression in ovarian cancer. The Ovachip was shown to be highly accurate in distinguishing between ovarian cancer and colon cancer.

17.1.6 The Novartis Group

Welsh et al. (2001b) published a study of 27 serous papillary adenocarcinomas of the ovary and three normal ovarian samples using an Affymetrix HuGeneFL array wafer. Hierarchical clustering revealed subgroups consistent with known clinical features.

17.2 SUMMARY

The study of ovarian cancer with microarrays is dominated by small studies with few samples and few genes. They do agree, however, that it is possible to differentiate between metatstatic and nonmetastatic or between survival and nonsurvival. More studies with larger populations are needed. Several studies have also shown the ability of DNA microarrays to differentiate between ovarian cancer and colon, breast, and uterine cancer.

FURTHER READING

Haviv, I., and Campbell, I. G. (2002). DNA microarrays for assessing ovarian cancer gene expression. *Mol. Cell. Endocrinol.* 191(1):121–126 (review).

Hibbs, K., Skubitz, K. M., Pambuccian, S. E., Casey, R. C., Burleson, K. M., Oegema, T. R., Thiele, J. J., Grindle, S. M., Bliss, R. L., and Skubitz, A. P. (2004). Differential gene expression in ovarian carcinoma: identification of potential biomarkers. *Am. J. Pathol.* 165(2):397–414.

Hough, C. D., Cho, K. R., Zonderman, A. B., Schwartz, D. R., and Morin, P. J. (2001). Coordinately up-regulated genes in ovarian cancer. *Cancer Res.* **61**(10):3869–3876.

Ismail, R. S., Baldwin, R. L., Fang, J., Browning, D., Karlan, B. Y., Gasson, J. C., and Chang, D. D. (2000). Differential gene expression between normal and tumor-derived ovarian epithelial cells. *Cancer Res.* 60(23):6744–6749.

Lee, B. C., Cha, K., Avraham, S., and Avraham, H. K. (2004). Microarray analysis of differentially expressed genes associated with human ovarian cancer. *Int. J. Oncol.* 24(4):847–851.

Martoglio, A. M., Tom, B. D., Starkey, M., Corps, A. N., Charnock-Jones, D. S., and Smith, S. K. (2000). Changes in tumorigenesis- and angiogenesis-related gene transcript abundance profiles in ovarian cancer detected by tailored high density cDNA arrays. *Mol. Med.* 6(9):750–765.

Nishizuka, S., Chen, S. T., Gwadry, F. G., Alexander, J., Major, S. M., Scherf, U., Reinhold, W. C., Waltham, M., Charboneau, L., Young, L., Bussey, K. J., Kim, S., Lababidi, S., Lee, J. K., Pittaluga, S., Scudiero, D. A., Sausville, E. A., Munson, P. J., Petricoin, E. F., Liotta, L. A., Hewitt, S. M., Raffeld, M., and Weinstein, J. N. (2003). Diagnostic markers that distinguish colon and ovarian adenocarcinomas: identification by genomic, proteomic, and tissue array profiling. *Cancer Res.* 63(17):5243–5250.

Presneau, N., Mes-Masson, A. M., Ge, B., Provencher, D., Hudson, T. J., and Tonin, P. N. (2003). Patterns of expression of chromosome 17 genes in primary cultures of normal ovarian surface epithelia and epithelial ovarian cancer cell lines. *Oncogene* 22(10):1568–1579.

Sakamoto, M., Kondo, A., Kawasaki, K., Goto, T., Sakamoto, H., Miyake, K., Koyamatsu, Y., Akiya, T., Iwabuchi, H., Muroya, T., Ochiai, K., Tanaka, T., Kikuchi, Y., and Tenjin, Y. (2001). Analysis of gene expression profiles associated with cisplatin resistance in human ovarian cancer cell lines and tissues using cDNA microarray. *Hum. Cell* 14(4):305–315.

Santin, A. D., Zhan, F., Bellone, S., Palmieri, M., Cane, S., Gokden, M., Roman, J. J., O'Brien, T. J., Tian, E., Cannon, M. J., Shaughnessy, J., and Pecorelli, S. (2004). Discrimination between uterine serous papillary carcinomas and ovarian serous papillary tumours by gene expression profiling. *Br. J. Cancer* 90(9):1814–1824.

Santin, A. D., Zhan, F., Bellone, S., Palmieri, M., Cane, S., Bignotti, E., Anfossi, S., Gokden, M., Dunn, D., Roman, J. J., O'Brien, T. J., Tian, E., Cannon, M. J., Shaughnessy, J., and Pecorelli, S. (2004). Gene expression profiles in primary ovarian serous papillary tumors and normal ovarian epithelium: identification of candidate molecular markers for ovarian cancer diagnosis and therapy. *Int. J. Cancer* 112(1):14–25.

Schwartz, D. R., Kardia, S. L., Shedden, K. A., Kuick, R., Michailidis, G., Taylor, J. M., Misek, D. E., Wu, R., Zhai, Y., Darrah, D. M., Reed, H., Ellenson, L. H., Giordano, T. J., Fearon, E. R., Hanash, S. M., and Cho, K. R. (2002). Gene expression in ovarian cancer reflects both morphology and biological behavior, distinguishing clear cell from other poor-prognosis ovarian carcinomas. *Cancer Res.* 62(16):4722–4729.

Spentzos, D., Levine, D. A., Kolia, S., Otu, H., Boyd, J., Libermann, T. A., and Cannistra, S. A. (2005). Unique gene expression profile based on pathologic response in epithelial ovarian cancer. *J. Clin. Oncol.* 23(31):7911–7918.

Tapper, J., Kettunen, E., El-Rifai, W., Seppala, M., Andersson, L. C., and Knuutila, S. (2001). Changes in gene expression during progression of ovarian carcinoma. *Cancer Genet. Cytogenet.* 128(1):1–6.

Tonin, P. N., Hudson, T. J., Rodier, F., Bossolasco, M., Lee, P. D., Novak, J., Manderson, E. N., Provencher, D., and Mes-Masson, A. M. (2001). Microarray analysis of gene expression mirrors the biology of an ovarian cancer model. *Oncogene* 20(45):6617–6626.

Wang, K., Gan, L., Jeffery, E., Gayle, M., Gown, A. M., Skelly, M., Nelson, P. S., Ng, W. V., Schummer, M., Hood, L., and Mulligan, J. (1999). Monitoring gene expression profile changes in ovarian carcinomas using cDNA microarray. *Gene* 229(1-2):101–108.

Wong, K. K., Cheng, R. S., and Mok, S. C. (2001). Identification of differentially expressed genes from ovarian cancer cells by MICROMAX cDNA microarray system. *Biotechniques* 30(3):670–675.

Xu, S., Mou, H., Lu, G., Zhu, C., Yang, Z., Gao, Y., Lou, H., Liu, X., Cheng, Y., and Yang, W. (2002). Gene expression profile differences in high and low metastatic human ovarian cancer cell lines by gene chip. *Chin. Med. J. (Engl.)* 115(1):36–41.

18

Prostate Cancer

The prostate is found only in men and contains gland cells that produce some of the seminal fluid that protects and nourishes sperm cells (Figure 18.1). More than 99% of prostate cancers are adenocarcinomas, referring to their origin in the gland cells.

Prostate cancer is diagnosed by taking one or more biopsies through the rectum.

Pathologists grade prostate cancer on the Gleason scale from 1 to 5, depending on how much the cells from the biopsy, when examined under a microscope, look like normal cells. Because of heterogeneity of prostate cancers, Gleason grades from two different biopsies are often added together to form a Gleason score (Gleason sum) between 2 and 10. In addition, the TNM staging system of the American Joint Committee on Cancer (AJCC) is also used on prostate cancer.

The main treatment types for prostate cancer are surgery, radiation therapy, chemotherapy, and hormone therapy.

18.1 MICROARRAY STUDIES OF PROSTATE CANCER

18.1.1 University of Michigan Medical School

Dhanasekaran et al. (2001) published a *Nature* paper on the study of 50 normal and neoplastic prostate specimens using cDNA microarrays. Signature expression profiles of normal adjacent prostate (NAP), BPH, localized prostate cancer, and metastatic, hormone-refractory prostate cancer were determined. The expression of two proteins was assessed by a tissue microarray with 700 clinically stratified prostate cancer specimens and showed significant correlation with measures of outcome.

18.1.2 Memorial Sloan-Kettering Cancer Center

LaTulippe et al. (2002) published a study of 23 primary prostate cancers, 9 metastatic prostate cancers, and 3 normal prostates using the Affymetrix U95 GeneChip series.

Cancer Diagnostics with DNA Microarrays, By Steen Knudsen
Copyright © 2006 John Wiley & Sons, Inc.

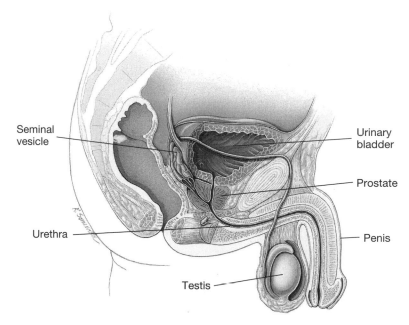

Figure 18.1 *The prostate and the male organs of reproduction. (From Tortora, Principles of Human Anatomy, 10th ed., 2005, p. 836. Used with permission of John Wiley & Sons, Inc.)*

The mean expression level of more than 300 genes differed more than threefold between recurrent and nonrecurrent cancers. No statistical analysis was performed.

18.1.3 The NIH Group

Best et al. (2003) published a study of 13 high- and moderate-grade human prostate tumors using cDNA microarrays. The expression of 136 genes was found to differ significantly between normal prostate and tumors. Twenty-one genes differed between high- and moderate-grade tumors. A meta-analysis of four different datasets confirmed some of the findings of their study.

18.1.4 The Harvard Group

Sing et al. (2002) published a study of 52 prostate tumors and 50 normal prostate samples using Affymetrix U95Av2 GeneChips. A set of genes was found that strongly correlated with the state of tumor differentiation as measured by the Gleason score. A KNN classifier that could predict patient outcome following prostatectomy with 90% accuracy was built based on 5 genes.

18.1.5 The Australian Group

Henshall et al. (2003) published a study of 72 prostate cancers using a custom Affymetrix GeneChip with 59,619 probe sets. They found 266 probe sets with strong correlations to clinical outcome (relapse), confirming some of the earlier findings by other groups.

18.1.6 The Stanford Group

Lapointe et al. (2004) published a study of 62 primary prostate tumors, 41 normal prostate specimens, and 9 lymph node metastases using cDNA microarrays containing approximately 26,000 genes. Unsupervised hierarchical clustering identified subclasses with differing aggressiveness.

18.2 SUMMARY

Several studies have shown that prostate cancer can be diagnosed and a relevant prognosis determined using DNA microarrays. DNA microarrays appear ready to be deployed for the diagnosis of prostate cancer. At least 384 patients have been studied so far.

FURTHER READING

Bull, J. H., Ellison, G., Patel, A., Muir, G., Walker, M., Underwood, M., Khan, F., and Paskins, L. (2001). Identification of potential diagnostic markers of prostate cancer and prostatic intraepithelial neoplasia using cDNA microarray. *Br. J. Cancer* 84(11):1512–1519.

Elek, J., Park, K. H., and Narayanan, R. (2000). Microarray-based expression profiling in prostate tumors. *In Vivo* 14(1):173–182.

Ernst, T., Hergenhahn, M., Kenzelmann, M., Cohen, C. D., Bonrouhi, M., Weninger, A., Klaren, R., Grone, E. F., Wiesel, M., Gudemann, C., Kuster, J., Schott, W., Staehler, G., Kretzler, M., Hollstein, M., and Grone, H. J. (2002). Decrease and gain of gene expression are equally discriminatory markers for prostate carcinoma: a gene expression analysis on total and microdissected prostate tissue. *Am. J. Pathol.* 160(6):2169–2180.

Febbo, P. G., and Sellers, W. R. (2003). Use of expression analysis to predict outcome after radical prostatectomy. *J. Urol.* 170(6 Pt 2):S11–S19.

Huppi, K., and Chandramouli, G. V. (2004). Molecular profiling of prostate cancer. *Curr. Urol. Rep.* 5(1):45–51. (review).

Kristiansen, G., Pilarsky, C., Wissmann, C., Kaiser, S., Bruemmendorf, T., Roepcke, S., Dahl, E., Hinzmann, B., Specht, T., Pervan, J., Stephan, C., Loening, S., Dietel, M., and Rosenthal, A. (2005). Expression profiling of microdissected matched prostate cancer samples reveals CD166/MEMD and CD24 as new prognostic markers for patient survival. *J. Pathol.* 205(3):359–376.

Latil, A., Bieche, I., Chene, L., Laurendeau, I., Berthon, P., Cussenot, O., and Vidaud, M. (2003). Gene expression profiling in clinically localized prostate cancer: a four-gene expression model predicts clinical behavior. *Clin. Cancer Res.* 9(15):5477–5485.

Luo, J. H., Yu, Y. P., Cieply, K., Lin, F., Deflavia, P., Dhir, R., Finkelstein, S., Michalopoulos, G., and Becich, M. (2002). Gene expression analysis of prostate cancers. *Mol. Carcinog.* 33(1):25–35.

Magee, J. A., Araki, T., Patil, S., Ehrig, T., True, L., Humphrey, P. A., Catalona, W. J., Watson, M. A., and Milbrandt, J. (2001). Expression profiling reveals hepsin overexpression in prostate cancer. *Cancer Res.* 61(15):5692–5696.

Stuart, R. O., Wachsman, W., Berry, C. C., Wang-Rodriguez, J., Wasserman, L., Klacansky, I., Masys, D., Arden, K., Goodison, S., McClelland, M., Wang, Y., Sawyers, A., Kalcheva, I., Tarin, D., and Mercola, D. In silico dissection of cell-type-associated patterns of gene expression in prostate cancer. *Proc. Natl. Acad. Sci. USA* 101(2):615–620.

Welsh, J. B., Sapinoso, L. M., Su, A. I., Kern, S. G., Wang-Rodriguez, J., Moskaluk, C. A., Frierson, H. F., and Hampton, G. M. (2001). Analysis of gene expression identifies candidate markers and pharmacological targets in prostate cancer. *Cancer Res.* 61(16):5974–5978.

Xu, J., Stolk, J. A., Zhang, X., Silva, S. J., Houghton, R. L., Matsumura, M., Vedvick, T. S., Leslie, K. B., Badaro, R., and Reed, S. G. (2000). Identification of differentially expressed genes in human prostate cancer using subtraction and microarray. *Cancer Res.* 60(6):1677–1682.

19

Melanoma

Melanoma is a cancer that originates in the melanocytes, the melanin producing cells in the epidermis (Figure 19.1). Melanomas have a high potential for metastasis. Uveal melanoma is a melanoma that originates in the eye.

Cancers of the skin account for more than 50% of all cancers (American Cancer Society, 2004). Of these, 4% are melanomas, but they account for 79% of skin cancer deaths, illustrating the aggressive nature of the melanomas.

Diagnosis of melanoma is by tissue biopsy, and diagnosis of metastasis is typically by fine needle aspiration from a metastasis, for example, to a lymph node.

Current treatment of melanoma includes surgery, radiation therapy, chemotherapy, and immunotherapy.

19.1 MICROARRAY STUDIES OF MELANOMA

19.1.1 Memorial Sloan-Kettering Cancer Center

Segal et al. (2003) published a study of 21 cell lines and 60 melanoma samples using Affymetrix U95A GeneChips. Unsupervised clustering showed a clear separation between melanoma and soft tissue sarcoma (STS). Furthermore, clear cell sarcoma (CCS) clustered as a separate group within the melanomas. A support vector machine was built to diagnose melanomas.

19.1.2 The NIH Group

Bittner et al. (2000) published a study of 31 melanomas and 7 controls analyzed with a cDNA array with 8150 clones. Using hierarchical cluster analysis and multidimensional scaling, they identified a new subgroup of 19 melanomas. This group has a better prognosis than the rest, though the difference was not statistically significant due to the low numbers.

Cancer Diagnostics with DNA Microarrays, By Steen Knudsen
Copyright © 2006 John Wiley & Sons, Inc.

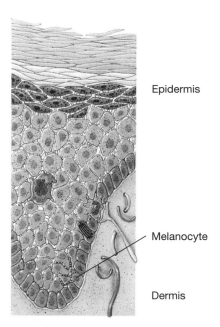

Epidermis

Melanocyte

Dermis

Figure 19.1 *Melanocytes in the epidermis of the skin. (From Tortora,* Principles of Human Anatomy, *10th ed., 2005, p. 124. Used with permission of John Wiley & Sons, Inc.)*

Wang et al. (2002) published a study of 63 fine needle aspiration samples from 37 melanoma metastases from 25 patients undergoing immunotherapy. A 6108-gene human cDNA chip was used. They identified 30 genes that correlated with clinical outcome (response to immunotherapy) prospectively.

19.2 SUMMARY

Most studies of melanoma have been with cell lines and mouse models looking for markers of metastasis or targets for drug development. But a few studies have conducted transcription profiling of larger patient cohorts and showed that microarrays can be used for diagnosis and prognosis of melanomas and their metastasis. More studies are needed to confirm these observations.

FURTHER READING

Baldi, A., Battista, T., De, L. U., Santini, D., Rossiello, L., Baldi, F., Natali, P. G., Lombardi, D., Picardo, M., Felsani, A., and Paggi, M. G. (2003). Identification of genes down-regulated during melanoma progression: a cDNA array study. *Exp. Dermatol.* 12(2):213–218.

Baldi, A., Santini, D., De, L. U., and Paggi, M. G. (2003). cDNA array technology in melanoma: an overview. *J. Cell. Physiol.* 196(2):219–223 (review).

Becker, B., Roesch, A., Hafner, C., Stolz, W., Dugas, M., Landthaler, M., and Vogt, T. (2004). Discrimination of melanocytic tumors by cDNA array hybridization of tissues prepared by laser pressure catapulting. *J. Invest. Dermatol.* 122(2):361–368.

Brem, R., Hildebrandt, T., Jarsch, M., Van Muijen, G. U., and Weidle, U. H. (2001). Identification of metastasis-associated genes by transcriptional profiling of a metastasizing versus a non-metastasizing human melanoma cell line. *Anticancer Res.* 21(3B):1731–1740.

Carr, K. M., Bittner, M., Trent, J. M. (2003). Gene-expression profiling in human cutaneous melanoma. *Oncogene* 22(20):3076–3080 (review).

Clark, E. A., Golub, T. R., Lander, E. S., and Hynes, R. O. (2000). Genomic analysis of metastasis reveals an essential role for RhoC. *Nature* 406(6795):532–535. Erratum in: *Nature* 2001; 411(6840):974.

Conway, R. M., Cursiefen, C., Behrens, J., Naumann, G. O., and Holbach, L. M. (2003). Biomolecular markers of malignancy in human uveal melanoma: the role of the cadherin–catenin complex and gene expression profiling. *Ophthalmologica* 217(1):68-75 (review).

de Wit, N. J., Burtscher, H. J., Weidle, U. H., Ruiter, D. J., and van Muijen, G. N. (2002). Differentially expressed genes identified in human melanoma cell lines with different metastatic behaviour using high density oligonucleotide arrays. *Melanoma Res.* 12(1):57–69.

Dooley, T. P., Reddy, S. P., Wilborn, T. W., and Davis, R. L. (2003). Biomarkers of human cutaneous squamous cell carcinoma from tissues and cell lines identified by DNA microarrays and qRT-PCR. *Biochem. Biophys. Res. Commun.* 306(4):1026–1036.

Kim, C. J., Reintgen, D. S., and Yeatman, T. J. (2002). The promise of microarray technology in melanoma care. *Cancer Control* 9(1):49–53 (review).

Pavey, S., Johansson, P., Packer, L., Taylor, J., Stark, M., Pollock, P. M., Walker, G. J., Boyle, G. M., Harper, U., Cozzi, S. J., Hansen, K., Yudt, L., Schmidt, C., Hersey, P., Ellem, K. A., O'Rourke, M. G., Parsons, P. G., Meltzer, P., Ringner, M., and Hayward, N. K. (2004). Microarray expression profiling in melanoma reveals a BRAF mutation signature. *Oncogene* 23(23):4060–4067.

McDonald, S. L., Edington, H. D., Kirkwood, J. M., and Becker, D. (2004). Expression analysis of genes identified by molecular profiling of VGP melanomas and MGP melanoma-positive lymph nodes. *Cancer Biol. Ther.* 3(1):110–120.

Roesch, A., Vogt, T., Stolz, W., Dugas, M., Landthaler, M., and Becker, B. (2003). Discrimination between gene expression patterns in the invasive margin and the tumour core of malignant melanomas. *Melanoma Res.* 13(5):503–509.

Seftor, E. A., Meltzer, P. S., Kirschmann, D. A., Pe'er, J., Maniotis, A. J., Trent, J. M., Folberg, R., and Hendrix, M. J. (2002). Molecular determinants of human uveal melanoma invasion and metastasis. *Clin. Exp. Metastasis* 19(3):233–246.

Tschentscher, F., Husing, J., Holter, T., Kruse, E., Dresen, I. G., Jockel, K. H., Anastassiou, G., Schilling, H., Bornfeld, N., Horsthemke, B., Lohmann, D. R., and Zeschnigk, M. (2003). Tumor classification based on gene expression profiling shows that uveal melanomas with and without monosomy 3 represent two distinct entities. *Cancer Res.* 63(10):2578–2584.

van der Velden, P. A., Zuidervaart, W., Hurks, M. H., Pavey, S., Ksander, B. R., Krijgsman, E., Frants, R. R., Tensen, C. P., Willemze, R., Jager, M. J., and Gruis, N. A. (2003). Expression profiling reveals that methylation of TIMP3 is involved in uveal melanoma development. *Int. J. Cancer* 106(4):472–479.

Zuidervaart, W., van der Velden, P. A., Hurks, M. H., van Nieuwpoort, F. A., Out-Luiting, C. J., Singh, A. D., Frants, R. R., Jager, M. J., and Gruis, N. A. (2003). Gene expression profiling identifies tumour markers potentially playing a role in uveal melanoma development. *Br. J. Cancer* 89(10):1914–1919.

20

Brain Tumors

The central nervous system is constituted by the brain and the spinal cord. The brain consists of the cerebral hemispheres, the diencephalon, the cerebellum, and the brain stem (Figure 20.1). Brain and spinal cord tumors occur most often in children younger than 10 years. Brain tumors are characterized by the type of cell and region of the brain from which they originate. Some brain tumors originate in other parts of the body and metastasize to the brain. They are called metastatic brain tumors or secondary brain tumors to distinguish them from the primary brain tumors that originate in the brain.

Neurons are the most important cells (Figure 20.2) in the brain. Tumors arising from the neurons of the cerebellum are called medulloblastomas. There are four types of glial cells: astrocytes, oligodendrocytes, microglia, and ependymal cells. Tumors arising in glial cells are called gliomas, tumors arising in the astrocytes are called astrocytomas (high-grade astrocytoma is referred to as glioblastoma), tumors arising in the oligodendrocytes are called oligodendrogliomas, and tumors arising in the ependymal cells are called ependymomas.

In children, the most common brain tumor is supratentorial astrocytoma, accounting for 25–40% of all tumors (American Cancer Association, 2004). Supratentorial astrocytomas arise in astrocyte cells outside the cerebellum, brain stem, or spinal cord. The cerebellar astrocytomas account for 10–20% of all cases of brain tumors in children. The brain stem gliomas account for 10–20% of all childhood brain tumors. The medulloblastoma likewise accounts for about 10–20% of brain tumors. More rare forms of childhood brain tumors are ependymoma (5–10%), craniopharyngioma (6–9%), and pineal tumors (0.5–2%).

Diagnosis of brain tumors is typically by magnetic resonance imaging (MRI) and computer tomography (CT) scan, followed by a biopsy of a suspected tumor.

There is no generally accepted staging system for brain tumors. They are, however, graded into low, intermediate, and high grade based on the appearance of cells from

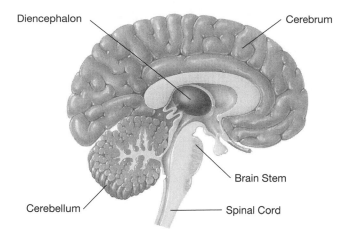

Diencephalon

Cerebrum

Brain Stem

Cerebellum

Spinal Cord

Figure 20.1 *Anatomy of the brain. (From Tortora, Principles of Human Anatomy, 10th ed., 2005, p. 585. Used with permission of John Wiley & Sons, Inc.)*

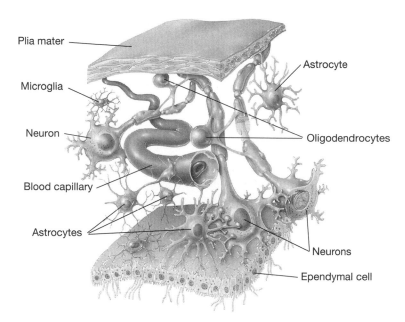

Plia mater

Astrocyte

Microglia

Neuron

Oligodendrocytes

Blood capillary

Astrocytes

Neurons

Ependymal cell

Figure 20.2 *Cells of the central nervous system. (From Tortora, Principles of Human Anatomy, 10th ed., 2005, p. 544. Used with permission of John Wiley & Sons, Inc.)*

a biopsy under the microscope. Medulloblastomas are graded according to the Chang system based on tumor size and spread (T) and evidence of metastasis to other parts of the brain or outside the brain (M).

Treatment of brain tumors consists of surgery, radiation therapy, and chemotherapy. More than half of all children with brain tumors survive five years after diagnosis.

20.1   MICROARRAY STUDIES OF BRAIN TUMORS

20.1.1   Harvard Medical School

Pomeroy et al. (2002) published a *Nature* paper on the study of 99 patient samples using Affymetrix HuGeneFL GeneChips. Principal component analysis and hierarchical clustering revealed that medulloblastomas are molecularly distinct from other brain tumors including primitive neuroectodermal tumors (PNETs), atypical teratoid/rhabdoid tumors (AT/RTs), and malignant gliomas. A classifier of tumor type was developed. Medulloblastoma outcome prediction was also possible using a KNN classifier with 8 genes that made only 13 classification errors on 60 samples. It was demonstrated that gene expression based outcome prediction substantially improved staging based prognostication.

In 2003 the group published a study of 50 gliomas using the Affymetrix U95Av2 GeneChip (Nutt et al., 2003). A classifier of prognosis was built that better correlates with clinical outcome than does standard pathology.

20.1.2   M. D. Anderson Cancer Center

Fuller et al. (2002) published a study of 30 primary human glioma tissue samples analyzed by cDNA arrays. Multidimensional scaling separated four different glioma subtypes: glioblastoma, anaplastic astrocytoma, anaplastic oligodendroglioma, and oligodendroglioma. Survival analysis revealed a good correlation to the molecular classification.

20.1.3   UCLA School of Medicine

Shai et al. (2003) published a study of gliomas of different types and grades using Affymetrix GeneChips. Unsupervised clustering revealed three groups corresponding to glioblastomas, lower grade astrocytomas, and oligodendrogliomas. A predictor constructed from 170 genes differentially expressed between the subsets correctly identified the type and grade of all samples.

In 2004 the group published a study of 74 gliomas on U133 GeneChips (Freije et al., 2004). A set of 44 genes correlated to survival was identified.

20.1.4   The Lausanne Group

Godard et al. (2003) published a study of 53 patient biopsies analyzed with cDNA microarrays. They found that different subtypes of glioma can be identified based on gene expression: low-grade astrocytoma, primary glioblastoma, and secondary glioblastoma. Based on these gene expression profiles, a nearly perfect glioma classifier was constructed.

20.1.5   Deutsches Krebsforschungszentrum

Neben et al. (2004) published a study of 35 medulloblastoma neoplasms. They found 54 genes that predicted unfavorable survival in medulloblastoma. Nine genes were confirmed with tissue microarrays containing 180 tumors.

20.1.6 Children's Hospital, Boston

Fernandez-Teijeiro et al. (2004) published a study of 55 young patients with medulloblastoma. Genes whose expression was correlated with patient outcome were analyzed in a Cox proportional hazards model and compared to other clinical variables. It was found that gene expression is the only significant prognostic factor. Clinical criteria did not significantly contribute additional information for outcome predictions.

20.2 SUMMARY

Several medium-sized clinical trials have shown that DNA microarrays can be used for diagnosis and prognosis of tumors of the brain. If these observations are confirmed in further studies, DNA microarrays appear ready to be deployed for the diagnosis of certain brain tumors. At least 346 patients have been studied so far.

FURTHER READING

Boon, K., Edwards, J. B., Siu, I. M., Olschner, D., Eberhart, C. G., Marra, M. A., Strausberg, R. L., and Riggins, G. J. (2003). Comparison of medulloblastoma and normal neural transcriptomes identifies a restricted set of activated genes. *Oncogene* **22**(48):7687–7694.

Huang, H., Colella, S., Kurrer, M., Yonekawa, Y., Kleihues, P., and Ohgaki, H. (2000). Gene expression profiling of low-grade diffuse astrocytomas by cDNA arrays. *Cancer Res.* **60**(24):6868–6874.

Khatua, S., Peterson, K. M., Brown, K. M., Lawlor, C., Santi, M. R., LaFleur, B., Dressman, D., Stephan, D. A., and MacDonald, T. J. (2003). Overexpression of the EGFR/FKBP12/HIF-2alpha pathway identified in childhood astrocytomas by angiogenesis gene profiling. *Cancer Res.* **63**(8):1865–1870.

Kim, S., Dougherty, E. R., Shmulevich, L., Hess, K. R., Hamilton, S. R., Trent, J. M., Fuller, G. N., and Zhang, W. (2002). Identification of combination gene sets for glioma classification. *Mol. Cancer Ther.* 1(13):1229–1236.

Korenberg, M. J. (2003). Gene expression monitoring accurately predicts medulloblastoma positive and negative clinical outcomes. *FEBS Lett.* 533(1-3):110–114.

Korshunov, A., Neben, K., Wrobel, G., Tews, B., Benner, A., Hahn, M., Golanov, A., and Lichter, P. (2003). Gene expression patterns in ependymomas correlate with tumor location, grade, and patient age. *Am. J. Pathol.* 163(5):1721–1727.

Ljubimova, J. Y., Khazenzon, N. M., Chen, Z., Neyman, Y. I., Turner, L., Riedinger, M. S., and Black, K. L. (2001). Gene expression abnormalities in human glial tumors identified by gene array. *Int. J. Oncol.* 18(2):287–295.

Mischel, P. S., Shai, R., Shi, T., Horvath, S., Lu, K. V., Choe, G., Seligson, D., Kremen, T. J., Palotie, A., Liau, L. M., Cloughesy, T. F., and Nelson, S. F. (2003). Identification of molecular subtypes of glioblastoma by gene expression profiling. *Oncogene* 22(15):2361–2373.

Park, P. C., Taylor, M. D., Mainprize, T. G., Becker, L. E., Ho, M., Dura, W. T., Squire, J., and Rutka, J. T. (2003). Transcriptional profiling of medulloblastoma in children. *J. Neurosurg.* 99(3):534–541.

Rickman, D. S., Bobek, M. P., Misek, D. E., Kuick, R., Blaivas, M., Kurnit, D. M., Taylor, J., and Hanash, S. M. (2001). Distinctive molecular profiles of high-grade and low-grade gliomas based on oligonucleotide microarray analysis. *Cancer Res.* 61(18):6885–6891.

Sallinen, S. L., Sallinen, P. K., Haapasalo, H. K., Helin, H. J., Helen, P. T., Schraml, P., Kallioniemi, O. P., and Kononen, J. (2000). Identification of differentially expressed genes in human gliomas by DNA microarray and tissue chip techniques. *Cancer Res.* 60(23):6617–6622.

van den Boom, J., Wolter, M., Kuick, R., Misek, D. E., Youkilis, A. S., Wechsler, D. S., Sommer, C., Reifenberger, G., and Hanash, S. M. (2003). Characterization of gene expression profiles associated with glioma progression using oligonucleotide-based microarray analysis and real-time reverse transcription-polymerase chain reaction. *Am. J. Pathol.* 163(3):1033–1043.

Watson, M. A., Perry, A., Budhjara, V., Hicks, C., Shannon, W. D., and Rich, K. M. (2001). Gene expression profiling with oligonucleotide microarrays distinguishes World Health Organization grade of oligodendrogliomas. *Cancer Res.* 61(5):1825–1829.

Wei, J. S., Greer, B. T., Westermann, F., Steinberg, S. M., Son, C. G., Chen, Q. R., Whiteford, C. C., Bilke, S., Krasnoselsky, A. L., Cenacchi, N., Catchpoole, D., Berthold, F., Schwab, M., and Khan, J. (2004). Prediction of clinical outcome using gene expression profiling and artificial neural networks for patients with neuroblastoma. *Cancer Res.* 64(19):6883–6891.

21

Organ or Tissue Specific Classification

Another major anticipated benefit of these technologies is the establishment of organ- and tumor-specific profiles that, among other potential benefits, might assist with the diagnostic work-up of patients with metastatic cancer of unknown origin at the time of initial diagnosis. A comprehensive library of unique gene expression profiles of all of the major tumor types would permit a definitive diagnosis in the absence of pertinent clinical history, imaging studies, and/or surgical exploration, thus simplifying the diagnostic evaluation.

Giordano et al., 2001

Most of this book is devoted to the application of DNA microarrays toward the prediction of the outcome of a cancer patient after surgery. That is because DNA microarrays promise to move oncology a significant step forward with this application. But there is another classification task in oncology that is actually easier to perform and where the results are even better. That is the distinction of the organ of origin for a specific tumor. For a primary tumor that question is of no clinical interest. But there is always a risk that a metastatic secondary tumor, that originated in another tissue, is mistaken for a primary tumor, leaving the primary tumor in the other tissue or organ untreated. DNA microarrays can distinguish between primary and secondary tumors, because they can differentiate between tissue and organ specific gene expression.

The first demonstration of the ability of DNA microarrays to distinguish between different tissues was published by Ramaswamy et al. (2001). They collected 218 tumor samples, spanning 14 common tumor types, and studied them with Affymetrix HU6800 and Hu35KsubA GeneChips. A Support Vector Machine based classifier was able to assign 78% of tumors to the correct organ. That included 8 metastatic (secondary) tumors, of which 6 were correctly identified as originating in another organ.

I repeated the work of Ramaswamy in 2004 with a database of 1125 cancer patients measured with Affymetrix chips (unpublished). These cancers reside in different organs:

breast, lung, prostate, kidney, brain, esophagus, pancreas, blood, lymph, and ovaries. The database of 1125 cancer patients was cross-validated; that is, the organ of the cancer was predicted based on the most similar other sample (nearest neighbor). Similarity was defined as Euclidean distance between expression vectors. In leave-one-out cross-validation the prediction of the organ of the cancer was 99.7% correct, meaning that it was in agreement with the pathological finding. A potential source of bias, however, was that all samples were not processed identically. They came from different studies.

A more specific example is lung neoplasms, which commonly develop in patients previously treated for head and neck carcinomas. The derivation of these tumors, either as new primary lung cancers or as metastatic head and neck cancers, is difficult to establish based on clinical or histopathologic criteria since both are squamous cell carcinomas and have identical features under light microscopy. Talbot et al. (2005) studied 52 lung cancer patients, of which 21 were primary lung tumors and 31 were primary head and neck tumors using Affymetrix HG-U95Av2 GeneChips. A molecular classifier showed a 98% accuracy in separating primary lung tumors from secondary lung tumors originating in the head and neck.

Similarly, the adenocarcinomas of the lung, colon, and ovary share similar features. Giordano et al. (2001) developed a molecular classifier based on the Affymetrix HuGeneFL GeneChip. It correctly classified 152 out of 154 samples according to their organ of origin.

FURTHER READING

Gordon, G. J., Jensen, R. V., Hsiao, L. L., Gullans, S. R., Blumenstock, J. E., Ramaswamy, S., Richards, W. G., Sugarbaker, D. J., and Bueno, R. (2002). Translation of microarray data into clinically relevant cancer diagnostic tests using gene expression ratios in lung cancer and mesothelioma. *Cancer Res.* 62(17):4963–4967.

22

Sample Collection and Stability

In living cells, the mRNA content is determined by a balance of synthesis and degradation. When cells or tissue are removed from the body, the supply of nutrients ceases and mRNA synthesis stops. Degradation remains and will over time reduce the mRNA pool. In addition there may be induction of stress-specific genes as the blood flow ceases. This has to be taken into consideration when collecting samples for DNA microarray analysis. The most widespread method to stop degradation is flash-freezing in liquid nitrogen and subsequent storage at -70 to $-80\,^{\circ}$C. This method has several disadvantages, however. It requires liquid nitrogen available in the operation room and it requires dry ice for shipping. Finally, samples must be homogenized before thawing to prevent resumption of mRNA degradation.

A convenient alternative is storing the sample in an ammonium sulfate solution, which precipitates the degrading RNAse enzymes. Such a solution is commercially available from Ambion (RNAlater).

The questions that remain are these: How soon must the sample be transferred to a stable condition? How stable is RNA during storage? Several published studies have sought to answer these questions.

22.1 TISSUE SAMPLES

22.1.1 Stability of Tissue After Surgical Removal

Ohashi et al. (2004) conducted a time-course study to determine whether any significant mRNA degradation could be observed in human breast tissue samples after surgical removal and before transfer of the sample to the stabilizing solution RNAlater. RNA integrity was judged by agarose gel electrophoresis and by RT-PCR of four genes. No difference in mRNA integrity was observed during the time course from 10 minutes to 3 hours. The authors conclude that it is safe to transfer the tissue up to 3 hours after surgical removal. Fitzpatrick et al. (2002) found RNA stable in bovine reproductive tissues up to 24 hours postmortem.

Yasojima et al. (2001) found little deterioration in RNA in brain tissue up to 96 hours postmortem.

Blackhall et al. (2004) studied lung cancer samples and found similar gene expression profiles regardless of the time that had elapsed between resection and freezing. They studied a time interval up to 120 minutes.

22.1.2 Stability of Sample in Storage

Mutter et al. (2004) showed that a tissue sample is stable for at least 72 hours at room temperature in RNAlater, allowing it to be shipped without cooling. They also showed that samples kept at room temperature in RNAlater for 72 hours did not give significantly different results on an Affymetrix U133A GeneChip than samples analyzed immediately (fresh) or samples frozen in liquid nitrogen.

Grotzer et al. (2000) found the same result for samples stored 7 days at room temperature in RNAlater.

As for long-term storage, Roos-van Groningen et al. (2004) found RNA stable for up to 2 months at $-70\ ^{\circ}$C in phosphate-buffered saline (PBS), and up to 3 months at $4\ ^{\circ}$C in RNAlater.

RNA in snap frozen tissue was found stable for 15 years at $-70\ ^{\circ}$C (Yasojima et al., 2001).

22.1.3 Paraffin-Embedded Tissue Samples

Large tissue banks have been built with formalin-fixed paraffin-embedded tissue samples (FFPE). This is not the best way of storing mRNA, but if FFPE tissue is your only option, there are solutions. It appears that the main problem is that the transcripts in FFPE tissue are shorter than in fresh or frozen tissue samples. This is due to chemical fragmentation and modification of the mRNA. Arcturus together with Affymetrix has developed a solution to this problem. A special human array X3P is available. It is designed to focus on interrogating sequences located closer to the $3'$ end of the transcripts compared with standard GeneChip arrays. Thus the array is less sensitive to strand breaks in the mRNA.

22.2 BLOOD SAMPLES

Tsui et al. (2002) studied the stability of blood plasma and serum. Following standard EDTA treatment, mRNA levels were stable in plasma for 24 hours at $4\ ^{\circ}$C. Muller et al. (2004) demonstrated that untreated peripheral blood left at room temperature experiences a dramatic loss of intact RNA. A stabilizing solution added to the blood (PAXgene Blood RNA Kit) could stabilize the RNA for 3 days at room temperature as measured by quantitative real-time PCR.

22.3 SAMPLE HETEROGENEITY

Blackhall et al. (2004) found considerable heterogeneity in expression levels of stress and hypoxia-activated genes in samples obtained from different areas of a lung tumor specimen at one time point after resection. They suggest pooling from multiple sites to reduce the heterogeneity. It remains, however, to be shown whether the heterogeneity in expression levels affects the outcome of classification based on the activity of other genes. Francis et al. (2005) addressed this issue by comparing 8–10 samples from the same tumor to one sample obtained from 16–20 different tumors. Both for malignant fibrous histiocytoma and a leiomyosarcoma they found that samples from the same tumor were more alike than samples from different tumors in hierarchical clustering. Thus Francis et al. (2005) suggest that a single sample is representative of the tumor for classification purposes. Schmulevich et al. (2002) came to the same conclusion in a study of leiomyosarcoma tumor heterogeneity.

Numerous single-disease studies have also measured sample heterogeneity as part of the study and found that it is not a problem for classification purposes (e.g., see Beer et al., 2002).

22.4 SUMMARY

A stabilizing solution such as RNAlater from Ambion makes sample handling very easy. Shipping at room temperature in RNAlater is no problem. Note that in many of the above studies RNA integrity has been judged by PCR. While it does not follow that the RNA is also suitable for microarray analysis, PCR is a very good indication of RNA integrity.

FURTHER READING

Cronin, M., Pho, M., Dutta, D., Stephans, J. C., Shak, S., Kiefer, M. C., Esteban, J. M., and Baker, J. B. (2004). Measurement of gene expression in archival paraffin-embedded tissues: development and performance of a 92-gene reverse transcriptase-polymerase chain reaction assay. *Am. J. Pathol.* 164(1):35–42.

Kopreski, M. S., Benko, F. A., Kwak, L. W., and Gocke, C. D. (1999). Detection of tumor messenger RNA in the serum of patients with malignant melanoma. *Clin. Cancer Res.* 5(8):1961–1965.

Perlmutter, M. A., Best, C. J., Gillespie, J. W., Gathright, Y., Gonzalez, S., Velasco, A., Linehan, W. M., Emmert-Buck, M. R., and Chuaqui, R. F. (2004). Comparison of snap freezing versus ethanol fixation for gene expression profiling of tissue specimens. *J. Mol. Diagn.* 6(4):371–377.

References

Adib, T. R., Henderson, S., Perrett, C., Hewitt, D., Bourmpoulia, D., Ledermann, J., and Boshoff, C. (2004). Predicting biomarkers for ovarian cancer using gene-expression microarrays. *Br. J. Cancer* 90(3):686–692.

Aerts, S., Van Loo, P., Thijs, G., Moreau, Y., and De Moor, B. (2003). Computational detection of cis -regulatory modules. *Bioinformatics* 19(Suppl 2):II5–II14.

Affymetrix (1999). *GeneChip Analysis Suite.* User Guide, version 3.3.

Affymetrix (2000). *GeneChip Expression Analysis.* Technical Manual.

Ahrendt, S. A., Halachmi, S., Chow, J. T., Wu, L., Halachmi, N., Yang, S. C., Wehage, S., Jen, J., and Sidransky, D. (1999). Rapid p53 sequence analysis in primary lung cancer using an oligonucleotide probe array. *Proc. Natl. Acad. Sci. USA* 96:7382–7387.

Albert, T. J., Norton, J., Ott, M., Richmond, T., Nuwaysir, K., Nuwaysir, E. F., Stengele, K. P., and Green, R. D. (2003). Light-directed $5' \rightarrow 3'$ synthesis of complex oligonucleotide microarrays. *Nucleic Acids Res.* 31(7):e35.

Alizadeh, A. A., Eisen, M. B., Davis, R. E., Ma, C., Lossos, I. S., Rosenwald, A., Boldrick, J. C., Sabet, H., Tran, T., Yu, X., Powell, J. I., Yang, L., Marti, G. E., Moore, T., Hudson, J., Lu, L., Lewis, D. B., Tibshirani, R., Sherlock, G., Chan, W. C., Greiner, T. C., Weisenburger, D. D., Armitage, J. O., Warnke, R., Levy, R., Wilson, W., Grever, M. R., Byrd, J. C., Botstein, D., Brown, P. O., and Staudt, L. M. (2000). Distinct types of diffuse large B-cell lymphoma identified by gene expression profiling. *Nature* 403(6769):503–511.

Alon, U., Barkai, N., Notterman, D. A., Gish, K., Ybarra, S., Mack, D., and Levine, A. J. (1999). Broad patterns of gene expression revealed by clustering analysis of tumor and normal colon tissues probed by oligonucleotide arrays. *Proc. Natl. Acad. Sci. USA* 96(12):6745–6750.

Alter, O., Brown, P. O., and Botstein, D. (2000). Singular value decomposition for genome-wide expression data processing and modeling. *Proc. Natl. Acad. Sci. USA* 97:10101–10106.

Altschul, S. F., Gish, W., Miller, W., Myers, E. W., and Lipman, D. J. (1990). Basic local alignment search tool. *J. Mol. Biol.* 215:403–410. (Available at http://www.ncbi.nlm.nih.gov/BLAST/.)

Altschul, S. F., Madden, T. L., Schäffer, A. A., Zhang, J., Zhang, Z., Miller, W., and Lipman, D. J. (1997). Gapped BLAST and PSI-BLAST: a new generation of protein database search programs. *Nucleic Acids Res.* 25:3389–3402. (Available at http://www.ncbi.nlm.nih.gov/BLAST/.)

Antal, P., Fannes, G., Timmerman, D., Moreau, Y., and De Moor, B. (2003). Bayesian applications of belief networks and multilayer perceptrons for ovarian tumor classification with rejection. *Artif. Intell. Med.* 29(1-2):39–60.

Arango, D., Laiho, P., Kokko, A., Alhopuro, P., Sammalkorpi, H., Salovaara, R., Nicorici, D., Hautaniemi, S., Alazzouzi, H., Mecklin, J. P., Jarvinen, H., Hemminki, A., Astola, J., Schwartz, S., and Aaltonen, L. A. (2005). Gene-expression profiling predicts recurrence in Dukes' C colorectal cancer. *Gastroenterology* 129(3):874–884.

Armstrong, S. A., Staunton, J. E., Silverman, L. B., Pieters, R., den Boer, M. L., Minden, M. D., Sallan, S. E., Lander, E. S., Golub, T. R., and Korsmeyer, S. J. (2002). MLL translocations specify a distinct gene expression profile that distinguishes a unique leukemia. *Nature Genet.* 30(1):41–47.

Audic, S., and Claverie, J. M. (1997). The significance of digital gene expression profiles. *Genome Res.* 7:986–995.

Baggerly, K. A., Coombes, K. R., Hess, K. R., Stivers, D. N, Abruzzo, L. V., and Zhang, W. (2001). Identifying differentially expressed genes in cDNA microarray experiments. *J. Computat. Biol.* 8(6):639–659.

Baldi, A., Battista, T., De Luca, A., Santini, D., Rossiello, L., Baldi, F., Natali, P. G., Lombardi, D., Picardo, M., Felsani, A., and Paggi, M. G. (2003a). Identification of genes downregulated during melanoma progression: a cDNA array study. *Exp. Dermatol.* 12(2):213–218.

Baldi, A., Santini, D., De Luca, A., and Paggi, M. G. (2003b). cDNA array technology in melanoma: an overview. *J. Cell. Physiol.* 196(2):219–223 (review).

Baldi, P., and Long, A. D. (2001). A Bayesian framework for the analysis of microarray expression data: regularized *t*-test and statistical inferences of gene changes. *Bioinformatics* 17:509–519. (Accompanying web page at http://visitor.ics.uci.edu/genex/cybert/.)

Baugh, L. R., Hill, A. A., Brown, E. L., and Hunter, C. P. (2001). Quantitative analysis of mRNA amplification by in vitro transcription. *Nucleic Acids Res.* 29:e29.

Baum, M., Bielau, S., Rittner, N., Schmid, K., Eggelbusch, K., Dahms, M., Schlauersbach, A., Tahedl, H., Beier, M., Guimil, R., Scheffler, M., Hermann, C., Funk, J. M., Wixmerten, A., Rebscher, H., Honig, M., Andreae, C., Buchner, D., Moschel, E., Glathe, A., Jager, E., Thom, M., Greil, A., Bestvater, F., Obermeier, F., Burgmaier, J., Thome, K., Weichert, S., Hein, S., Binnewies, T., Foitzik, V., Muller, M., Stahler, C. F., and Stahler, P. F. (2003). Validation of a novel, fully integrated and flexible microarray benchtop facility for gene expression profiling. *Nucleic Acids Res.* 31(23):e151.

Becker, B., Roesch, A., Hafner, C., Stolz, W., Dugas, M., Landthaler, M., and Vogt, T. (2004). Discrimination of melanocytic tumors by cDNA array hybridization of tissues prepared by laser pressure catapulting. *J. Invest. Dermatol.* 122(2):361–368.

Beer, D. G., Kardia, S. L., Huang, C. C., Giordano, T. J., Levin, A. M., Misek, D. E., Lin, L., Chen, G., Gharib, T. G., Thomas, D. G., Lizyness, M. L., Kuick, R., Hayasaka, S., Taylor, J. M., Iannettoni, M. D., Orringer, M. B., and Hanash, S. (2002). Gene-expression profiles predict survival of patients with lung adenocarcinoma. *Nature Med.* 8(8):816–824.

Beier, M., and Hoheisel, J. D. (2002). Analysis of DNA-microarrays produced by inverse in situ oligonucleotide synthesis. *J. Biotechnol.* 94:15–22.

Beier, M., Baum, M., Rebscher, H., Mauritz, R., Wixmerten, A., Stahler, C. F., Muller, M., and Stahler, P. F. (2002). Exploring nature's plasticity with a flexible probing tool, and finding new ways for its electronic distribution. *Biochem. Soc. Trans.* 30:78–82.

Bender, R., and Lange, S. (2001). Adjusting for multiple testing—when and how? *J. Clin. Epidemiol.* 54:343–349.

Ben-Hur, A., Elisseeff, A., and Guyon, I. (2002). A stability based method for discovering structure in clustered data. *Pacific Symposium on Biocomputing* 2002:6–17. (Available online at http://psb.stanford.edu.)

Benjamini, Y., and Hochberg, Y. (1995). Controlling the false discovery rate: a practical and powerful approach to multiple testing. *J. R. Statist. Soc. B* 57(1):289–300.

Bertucci, F., Borie, N., Ginestier, C., Groulet, A., Charafe-Jauffret, E., Adelaide, J., Geneix, J., Bachelart, L., Finetti, P., Koki, A., Hermitte, F., Hassoun, J., Debono, S., Viens, P., Fert, V., Jacquemier, J., and Birnbaum, D. (2004a). Identification and validation of an ERBB2 gene expression signature in breast cancers. *Oncogene* 23(14):2564–2575.

Bertucci, F., Salas, S., Eysteries, S., Nasser, V., Finetti, P., Ginestier, C., Charafe-Jauffret, E., Loriod, B., Bachelart, L., Montfort, J., Victorero, G., Viret, F., Ollendorff, V., Fert, V., Giovaninni, M., Delpero, J. R., Nguyen, C., Viens, P., Monges, G., Birnbaum, D., and Houlgatte, R. (2004b). Gene expression profiling of colon cancer by DNA microarrays and correlation with histoclinical parameters. *Oncogene* 23(7):1377–1391.

Best, C. J., Leiva, I. M., Chuaqui, R. F., Gillespie, J. W., Duray, P. H., Murgai, M., Zhao, Y., Simon, R., Kang, J. J., Green, J. E., Bostwick, D. G., Linehan, W. M., and Emmert-Buck, M. R. (2003). Molecular differentiation of high- and moderate-grade human prostate cancer by cDNA microarray analysis. *Diagn. Mol. Pathol.* 12(2):63–70.

Bhattacharjee, A., Richards, W. G., Staunton, J., Li, C., Monti, S., Vasa, P., Ladd, C., Beheshti, J., Bueno, R., Gillette, M., Loda, M., Weber, G., Mark, E. J., Lander, E. S., Wong, W., Johnson, B. E., Golub, T. R., Sugarbaker, D. J., and Meyerson, M. (2001). Classification of human lung carcinomas by mRNA expression profiling reveals distinct adenocarcinoma subclasses. *Proc. Natl. Acad. Sci. USA* 98(24):13790–13795.

Bicciato, S., Pandin, M., Didone, G., and Di Bello, C. (2003). Pattern identification and classification in gene expression data using an autoassociative neural network model. *Biotechnol. Bioeng.* 81(5):594–606.

Birkenkamp-Demtroder, K., Olesen, S. H., Sorensen, F. B., Laurberg, S., Laiho, P., Aaltonen, L. A., and Orntoft, T. F. (2005). Differential gene expression in colon cancer of the caecum versus the sigmoid and rectosigmoid. *Gut* 54(3):374–384.

Birnbaum, K., Benfey, P. N., and Shasha, D. E. (2001). Cis element/transcription factor analysis (cis/TF): a method for discovering transcription factor/cis element relationships. *Genome Res.* 11:1567–1573.

Bittner, M., Meltzer, P., Chen, Y., Jiang, Y., Seftor, E., Hendrix, M., Radmacher, M., Simon, R., Yakhini, Z., Ben-Dor, A., Sampas, N., Dougherty, E., Wang, E., Marincola, F., Gooden, C., Lueders, J., Glatfelter, A., Pollock, P., Carpten, J., Gillanders, E., Leja, D., Dietrich, K., Beaudry, C., Berens, M., Alberts, D., and Sondak, V. (2000). Molecular classification of cutaneous malignant melanoma by gene expression profiling. *Nature* 406(6795):536–540.

Black, M. A., and Doerge, R. W. (2002). Calculation of the minimum number of replicate spots required for detection of significant gene expression fold change in microarray experiments. *Bioinformatics* 18(12):1609–16.

Blackhall, F. H., Pintilie, M., Wigle, D. A., Jurisica, I., Liu, N., Radulovich, N., Johnston, M. R., Keshavjee, S., and Tsao, M. S. (2004). Stability and heterogeneity of expression profiles in lung cancer specimens harvested following surgical resection. *Neoplasia* 6(6):761–767.

Blamey, R. W., Davies, C. J., Elston, C. W., Johnson, J., Haybittle, J. L., and Maynard, P. V. (1979). Prognostic factors in breast cancer—the formation of a prognostic index. *Clin. Oncol.* 5(3):227–236.

Blaveri, E., Brewer, J. L., Roydasgupta, R., Fridlyand, J., Devries, S., Koppie, T., Pejavar, S., Mehta, K., Carroll, P., Simko, J. P., and Waldman, F. M. (2005a). Bladder cancer stage and outcome by array-based comparative genomic hybridization. *Clin. Cancer Res.* 11(19):7012–7022.

Blaveri, E., Simko, J. P., Korkola, J. E., Brewer, J. L., Baehner, F., Mehta, K., Devries, S., Koppie, T., Pejavar, S., Carroll, P., and Waldman, F. M. (2005b). Bladder cancer outcome and subtype classification by gene expression. *Clin. Cancer Res.* 11(11):4044–4055.

Bolstad, B. M., Irizarry, R. A., Astrand, M., and Speed, T. P. (2003). A comparison of normalization methods for high density oligonucleotide array data based on variance and bias. *Bioinformatics* 19(2):185–193.

Boon, K., Edwards, J. B., Siu, I. M., Olschner, D., Eberhart, C. G., Marra, M. A., Strausberg, R. L., and Riggins, G. J. (2003). Comparison of medulloblastoma and normal neural transcriptomes identifies a restricted set of activated genes. *Oncogene* 22(48):7687–7694.

Borczuk, A. C., Shah, L., Pearson, G. D., Walter, K. L., Wang, L., Austin, J. H., Friedman, R. A., and Powell, C. A. (2004). Molecular signatures in biopsy specimens of lung cancer. *Am. J. Respir. Crit. Care Med.* 170(2):167–174.

Borodovsky, M., and McIninch, J. (1993). GeneMark: parallel gene recognition for both DNA strands. *Computers & Chemistry* 17:123–133.

Bowtell, D., and Sambrook, J. (editors). (2002). *DNA Microarrays: A Molecular Cloning Manual.* New York: Cold Spring Harbor Laboratory Press.

Brazma, A., Jonassen, I., Vilo, J., and Ukkonen, E. (1998). Predicting gene regulatory elements in silico on a genomic scale. *Genome Res.* 8:1202–1215.

Brem, R., Hildebrandt, T., Jarsch, M., Van Muijen, G. U., and Weidle, U. H. (2001). Identification of metastasis-associated genes by transcriptional profiling of a metastasizing versus a non-metastasizing human melanoma cell line. *Anticancer Res.* 21(3B):1731–1740.

Brenner, S., Johnson, M., Bridgham, J., Golda, G., Lloyd, D. H., Johnson, D., Luo, S., McCurdy, S., Foy, M., Ewan, M., Roth, R., George, D., Eletr, S., Albrecht, G., Vermaas, E., Williams, S. R., Moon, K., Burcham, T., Pallas, M., DuBridge, R. B., Kirchner, J., Fearon, K., Mao, J., and Corcoran, K. (2000). Gene expression analysis by massively parallel signature sequencing (MPSS) on microbead arrays. *Nature Biotechnol.* 18:630–634.

Brown, M. P. S., Grundy, W. N., Lin, D., Cristianini, N., Sugnet, C. W., Furey, T. S., Ares, M., and Haussler, D. (2000). Knowledge-based analysis of microarray gene expression data by using support vector machines. *Proc. Natl. Acad. Sci. USA* 97:262–267.

Brunak, S., Engelbrecht, J., and Knudsen, S. (1990a). Cleaning up gene databases. *Nature* 343:123.

Brunak, S., Engelbrecht, J., and Knudsen, S. (1990b). Neural network detects errors in the assignment of mRNA splice sites. *Nucleic Acids Res.* 18:4797–4801.

Brunak, S., Engelbrecht, J., and Knudsen, S. (1991). Prediction of human mRNA donor and acceptor sites from the DNA sequence. *J. Mol. Biol.* 220:49–65.

Bull, J. H., Ellison, G., Patel, A., Muir, G., Walker, M., Underwood, M., Khan, F., and Paskins, L. (2001). Identification of potential diagnostic markers of prostate cancer and prostatic intraepithelial neoplasia using cDNA microarray. *Br. J. Cancer* 84(11):1512–1519.

Bullinger, L., Dohner, K., Bair, E., Frohling, S., Schlenk, R. F., Tibshirani, R., Dohner, H., and Pollack, J. R. (2004). Use of gene-expression profiling to identify prognostic subclasses in adult acute myeloid leukemia. *N. Engl. J. Med.* 350(16):1605–1616.

Burge C., and Karlin, S. (1997). Prediction of complete gene structures in human genomic DNA. *J. Mol. Biol.* 268:78–94.

Bussemaker, H. J., Li, H., and Siggia, E. D. (2000). Building a dictionary for genomes: identification of presumptive regulatory sites by statistical analysis. *Proc. Natl. Acad. Sci. USA* 97:10096–10100.

Cario, G., Stanulla, M., Fine, B. M., Teuffel, O., Neuhoff, N. V., Schrauder, A., Flohr, T., Schafer, B. W., Bartram, C. R., Welte, K., Schlegelberger, B., and Schrappe, M. (2005). Distinct gene expression profiles determine molecular treatment response in childhood acute lymphoblastic leukemia. *Blood* 105(2):821–826.

Carpentier, A. S., Riva, A., Tisseur, P., Didier, G., and Henaut, A. (2004). The operons, a criterion to compare the reliability of transcriptome analysis tools: ICA is more reliable than ANOVA, PLS and PCA. *Comput. Biol. Chem.* 28(1):3–10.

Carr, K. M., Bittner, M., Trent, J. M. (2003). Gene-expression profiling in human cutaneous melanoma. *Oncogene* 22(20):3076–3080 (review).

Carter, S. L., Eklund, A. C., Mecham, B. H., Kohane, I. S., and Szallasi, Z. (2005). Redefinition of Affymetrix probe sets by sequence overlap with cDNA microarray probes reduces cross-platform inconsistencies in cancer-associated gene expression measurements. *BMC Bioinformatics* 6(1):107.

Cheok, M. H., Yang, W., Pui, C. H., Downing, J. R., Cheng, C., Naeve, C. W., Relling, M. V., and Evans, W. E. (2003). Treatment-specific changes in gene expression discriminate in vivo drug response in human leukemia cells. *Nature Genetics* 34(1):85–90. Erratum in: *Nature Genet.* 2003 Jun;34(2):231. PMID: 12704389 [PubMed - indexed for MEDLINE].

Chiang, D. Y., Brown, P. O., and Eisen, M. B. (2001). Visualizing associations between genome sequences and gene expression data using genome-mean expression profiles. *Bioinformatics* 17(Suppl 1):S49–S55.

Chiappetta, P., Roubaud, M. C., and Torresani, B. (2004). Blind source separation and the analysis of microarray data. *J. Comput. Biol.* 11(6):1090–1109.

Choi, J. K., Yu, U., Kim, S., and Yoo, O. J. (2003). Combining multiple microarray studies and modeling interstudy variation. *Bioinformatics* 19 (Suppl 1):i84–i90.

Choi, J. K., Choi, J. Y., Kim, D. G., Choi, D. W., Kim, B. Y., Lee, K. H., Yeom, Y. I., Yoo, H. S., Yoo, O. J., and Kim, S. (2004). Integrative analysis of multiple gene expression profiles applied to liver cancer study. *FEBS Lett.* 565(1-3):93–100.

Clark, E. A., Golub, T. R., Lander, E. S., and Hynes, R. O. (2000). Genomic analysis of metastasis reveals an essential role for RhoC. *Nature* 406(6795):532–535. Erratum in: *Nature* 2000; 411(6840):974.

Claverie, J.-M. (1999). Computational methods for the identification of differential and coordinated gene expression. *Hum. Mol. Genet.* 8:1821–1832.

Cole, S. T., et al. (1998). Deciphering the biology of *Mycobacterium tuberculosis* from the complete genome sequence. *Nature* 393:537–544.

Cojocaru, G., Friedman, N., Krupsky, M., Yaron, P., Simansky, D., Yellin, A., Rechavi, G., Barash, Y., Ben-Dor, A., Yakhini, Z., and Kaminski, N. (2002). Transcriptional profiling of non-small cell lung cancer using oligonucleotide microarrays. *Chest* 121(3 Suppl):44S.

Conway, R. M., Cursiefen, C., Behrens, J., Naumann, G. O., and Holbach, L. M. (2003). Biomolecular markers of malignancy in human uveal melanoma: the role of the cadherin–catenin complex and gene expression profiling. *Ophthalmologica* 217(1):68-75 (review).

Cox, D. R., and Oakes, D. (1984). *Analysis of Survival Data.* New York: Chapman & Hall.

Dave, S. S., Wright, G., Tan, B., Rosenwald, A., Gascoyne, R. D., Chan, W. C., Fisher, R. I., Braziel, R. M., Rimsza, L. M., Grogan, T. M., Miller, T. P., LeBlanc, M., Greiner, T. C., Weisenburger, D. D., Lynch, J. C., Vose, J., Armitage, J. O., Smeland, E. B., Kvaloy, S., Holte, H., Delabie, J., Connors, J. M., Lansdorp, P. M., Ouyang, Q., Lister, T. A., Davies, A. J., Norton, A. J., Muller-Hermelink, H. K., Ott, G., Campo, E., Montserrat, E., Wilson, W. H., Jaffe, E. S., Simon, R., Yang, L., Powell, J., Zhao, H., Goldschmidt, N., Chiorazzi, M., and Staudt, L. M. (2004). Prediction of survival in follicular lymphoma based on molecular features of tumor-infiltrating immune cells. *N. Engl. J. Med.* 351(21):2159–2169.

Davis, R. E., and Staudt, L. M. (2002). Molecular diagnosis of lymphoid malignancies by gene expression profiling. *Curr. Opin. Hematol.* 9(4):333–338 (review).

De Smet, F., Mathys, J., Marchal, K., Thijs, G., De Moor, B., and Moreau, Y. (2002). Adaptive quality-based clustering of gene expression profiles. *Bioinformatics* 18(5):735–746.

de Wit, N. J., Burtscher, H. J., Weidle, U. H., Ruiter, D. J., and van Muijen, G. N. (2002). Differentially expressed genes identified in human melanoma cell lines with different metastatic behaviour using high density oligonucleotide arrays. *Melanoma Res.* 12(1):57–69.

Dhanasekaran, S. M., Barrette, T. R., Ghosh, D., Shah, R., Varambally, S., Kurachi, K., Pienta, K. J., Rubin, M. A., and Chinnaiyan, A. M. (2001). Delineation of prognostic biomarkers in prostate cancer. *Nature* 412(6849):822–826.

Dooley, T. P., Reddy, S. P., Wilborn, T. W., and Davis, R. L. (2003). Biomarkers of human cutaneous squamous cell carcinoma from tissues and cell lines identified by DNA microarrays and qRT-PCR. *Biochem. Biophys. Res. Commun.* 306(4):1026–1036.

Douglas, E. J., Fiegler, H., Rowan, A., Halford, S., Bicknell, D. C., Bodmer, W., Tomlinson, I. P., and Carter, N. P. (2004). Array comparative genomic hybridization analysis of colorectal cancer cell lines and primary carcinomas. *Cancer Res.* 64(14):4817–4825.

Dudoit, S., Fridlyand, J., and Speed, T. P. (2000a). Comparison of discrimination methods for the classification of tumors using gene expression data. Technical report #576, June 2000. (Available at http://www.stat.berkeley.edu/tech-reports/index.html.)

Dudoit, S., Yang, Y., Callow, M. J., and Speed, T. P. (2000b). Statistical methods for identifying differentially expressed genes in replicated cDNA microarray experiments. Technical report #578, August 2000. (Available at http://www.stat.berkeley.edu/tech-reports/index.html.)

Duggan, B. J., McKnight, J. J., Williamson, K. E., Loughrey, M., O'Rourke, D., Hamilton, P. W., Johnston, S. R., Schulman, C. C., and Zlotta, A. R. (2003). The need to embrace molecular profiling of tumor cells in prostate and bladder cancer. *Clin. Cancer Res.* 9(4):1240–1247 (review).

Durig, J., Nuckel, H., Huttmann, A., Kruse, E., Holter, T., Halfmeyer, K., Fuhrer, A., Rudolph, R., Kalhori, N., Nusch, A., Deaglio, S., Malavasi, F., Moroy, T., Klein-Hitpass, L., and Duhrsen, U. (2003). Expression of ribosomal and translation-associated genes is correlated with a favorable clinical course in chronic lymphocytic leukemia. *Blood* 101(7):2748–2755.

Dyrskjot, L. (2003). Classification of bladder cancer by microarray expression profiling: towards a general clinical use of microarrays in cancer diagnostics. *Expert Rev. Mol. Diagn.* 3(5):635–647. (review).

Dyrskjot, L., Thykjaer, T., Kruhoffer, M., Jensen, J. L., Marcussen, N., Hamilton-Dutoit, S., Wolf, H., and Orntoft, T. F. (2002). Identifying distinct classes of bladder carcinoma using microarrays. *Nature Genet.* 33(1):90–96.

Dyrskjot, L., Zieger, K., Kruhoffer, M., Thykjaer, T., Jensen, J. L., Primdahl, H., Aziz, N., Marcussen, N., Moller, K., and Orntoft, T. F. (2005). A molecular signature in superficial bladder carcinoma predicts clinical outcome. *Clin. Cancer Res.* 11(11):4029–4036.

Dysvik, B, and Jonassen, I. (2001). J-Express: exploring gene expression data using Java. *Bioinformatics* 17:369–370. (Software available at http://www.ii.uib.no/~bjarted/jexpress/.)

Eden, P., Ritz, C., Rose, C., Ferno, M., and Peterson, C. (2004). "Good old" clinical markers have similar power in breast cancer prognosis as microarray gene expression profilers. *Eur. J. Cancer* 40(12):1837–1841.

Efron, B., and Tibshirani, R. (2002). Empirical Bayes methods and false discovery rates for microarrays. *Genet. Epidemiol.* 23(1):70–86.

Efron, B., Storey, J., and Tibshirani, R. (2001). Microarrays, empirical Bayes methods, and false discovery rates. Technical report. Statistics Department, Stanford University. (Manuscript available at http://www-stat.stanford.edu/~tibs/research.html.)

Eifel, P., Axelson, J. A., Costa, J., Crowley, J., Curran, W. J., Deshler, A., Fulton, S., Hendricks, C. B., Kemeny, M., Kornblith, A. B., Louis, T. A., Markman, M., Mayer, R., and Roter, D. (2001). National Institutes of Health Consensus Development Conference Statement: adjuvant therapy for breast cancer, November 1–3, 2000. *J. Natl. Cancer Inst.* 93(13):979–989.

Elek, J., Park, K. H., and Narayanan, R. (2000). Microarray-based expression profiling in prostate tumors. *In Vivo* 14(1):173–182.

Ernst, T., Hergenhahn, M., Kenzelmann, M., Cohen, C. D., Bonrouhi, M., Weninger, A., Klaren, R., Grone, E. F., Wiesel, M., Gudemann, C., Kuster, J., Schott, W., Staehler, G., Kretzler, M., Hollstein, M., and Grone, H. J. (2002). Decrease and gain of gene expression are equally discriminatory markers for prostate carcinoma: a gene expression analysis on total and microdissected prostate tissue. *Am. J. Pathol.* 160(6):2169–2180.

Febbo, P. G., and Sellers, W. R. (2003). Use of expression analysis to predict outcome after radical prostatectomy. *J. Urol.* 170(6 Pt 2):S11–S19.

Fellenberg, K., Hauser, N. C., Brors, B., Neutzner, A., Hoheisel, J. D., and Vingron, M. (2001). Correspondence analysis applied to microarray data. *Proc. Natl. Acad. Sci. USA* 98:10781–10786.

Fernandez-Teijeiro, A., Betensky, R. A., Sturla, L. M., Kim, J. Y., Tamayo, P., and Pomeroy, S. L. (2004). Combining gene expression profiles and clinical parameters for risk stratification in medulloblastomas. *J. Clin. Oncol.* 22(6):994–998.

Ferrando, A. A., and Thomas Look, A. (2003). Gene expression profiling: will it complement or replace immunophenotyping? *Best Pract. Res. Clin. Haematol.* 16(4):645–652.

Ferrando, A. A., Neuberg, D. S., Staunton, J., Loh, M. L., Huard, C., Raimondi, S. C., Behm, F. G., Pui, C. H., Downing, J. R., Gilliland, D. G., Lander, E. S., Golub, T. R., and Look, A. T. (2002). Gene expression signatures define novel oncogenic pathways in T cell acute lymphoblastic leukemia. *Cancer Cell* 1(1):75–87.

Fine, B. M., Stanulla, M., Schrappe, M., Ho, M., Viehmann, S., Harbott, J., and Boxer, L. M. (2004). Gene expression patterns associated with recurrent chromosomal translocations in acute lymphoblastic leukemia. *Blood* 103(3):1043–1049. Epub 2003 Oct 02.

Fitzpatrick, R., Casey, O. M., Morris, D., Smith, T., Powell, R., and Sreenan, J. M.(2002). Postmortem stability of RNA isolated from bovine reproductive tissues. *Biochim. Biophys. Acta* 1574(1):10–14.

Fix, E., and Hodges, J. (1951). Discriminatory analysis, nonparametric discrimination: consistency properties. Technical report, Randolph Field, Texas: USAF School of Aviation Medicine.

Francis, P., Fernebro, J., Eden, P., Laurell, A., Rydholm, A., Domanski, H. A., Breslin, T., Hegardt, C., Borg, A., and Nilbert, M. (2005). Intratumor versus intertumor heterogeneity in gene expression profiles of soft-tissue sarcomas. *Genes Chromosomes Cancer* 43(3):302–308.

Frederiksen, C. M., Knudsen, S., Laurberg, S., and Orntoft, T. F. (2003). Classification of Dukes' B and C colorectal cancers using expression arrays. *J. Cancer Res. Clin. Oncol.* 129(5):263–271.

Freije, W. A., Castro-Vargas, F. E., Fang, Z., Horvath, S., Cloughesy, T., Liau, L. M., Mischel, P. S., and Nelson, S. F. (2004). Gene expression profiling of gliomas strongly predicts survival. *Cancer Res.* 64(18):6503–6510.

Fujibuchi, W., Anderson, J. S. J., and Landsman, D. (2001). PROSPECT improves cis-acting regulatory element prediction by integrating expression profile data with consensus pattern searches. *Nucleic Acids Res.* 29:3988–3996.

Fuller, G. N., Hess, K. R., Rhee, C. H., Yung, W. K., Sawaya, R. A., Bruner, J. M., and Zhang, W. (2002). Molecular classification of human diffuse gliomas by multidimensional scaling analysis of gene expression profiles parallels morphology-based classification, correlates with survival, and reveals clinically-relevant novel glioma subsets. *Brain Pathol.* 12(1):108–116 (review).

Garber, M. E., Troyanskaya, O. G., Schluens, K., Petersen, S., Thaesler, Z., Pacyna-Gengelbach, M., van de Rijn M., Rosen, G. D., Perou, C. M., Whyte, R. I., Altman, R. B., Brown, P. O., Botstein, D., and Petersen, I. (2001). Diversity of gene expression in adeno-carcinoma of the lung. *Proc. Natl. Acad. Sci. USA* 98(24):13784–13789. Erratum in: *Proc. Natl. Acad. Sci. USA* 2002; 99(2):1098.

Gautier, L., Cope, L., Bolstad B.M., and Irizarry R. A. (2004a). affy—Analysis of Affymetrix GeneChip data at the probe level. *Bioinformatics* 20(3):307–315.

Gautier, L, Møller, M., Friis-Hansen, L., and Knudsen S. (2004b). Alternative mapping of probes to genes for Affymetrix chips. *BMC Bioinformatics* 5(1):111.

Getz, G., Levine, E., and Domany, E. (2000). Coupled two-way clustering analysis of gene microarray data. *Proc. Natl. Acad. Sci. USA* 97:12079–12084.

Ghosh, D. (2002). Singular value decomposition regression models for classification of tumors from microarray experiments. *Pacific Symposium on Biocomputing* 2002:18–29. (Available online at http://psb.stanford.edu.)

Ghosh, D. (2003). Penalized discriminant methods for the classification of tumors from gene expression data. *Biometrics* 59(4):992–1000.

Ghosh, D., Barette, T. R., Rhodes, D., and Chinnaiyan, A. M. (2003). Statistical issues and methods for meta-analysis of microarray data: a case study in prostate cancer. *Funct. Integr. Genomics* 3(4):180–188.

Giles, P. J., and Kipling, D. (2003). Normality of oligonucleotide microarray data and implications for parametric statistical analyses. *Bioinformatics* 19:2254–2262 .

Giordano, T. J., Shedden, K. A., Schwartz, D. R., Kuick, R., Taylor, J. M., Lee, N., Misek, D. E., Greenson, J. K., Kardia, S. L., Beer, D. G., Rennert, G., Cho, K. R., Gruber, S. B., Fearon, E. R., and Hanash, S. (2001). Organ-specific molecular classification of primary lung, colon, and ovarian adenocarcinomas using gene expression profiles. *Am. J. Pathol.* 159(4):1231–1238.

Godard, S., Getz, G., Delorenzi, M., Farmer, P., Kobayashi, H., Desbaillets, I., Nozaki, M., Dis-erens, A. C., Hamou, M. F., Dietrich, P. Y., Regli, L., Janzer, R. C., Bucher, P., Stupp, R., de Tribolet N., Domany, E., and Hegi, M. E (2003). Classification of human astrocytic gliomas on the basis of gene expression: a correlated group of genes with angiogenic activity emerges as a strong predictor of subtypes. *Cancer Res.* 63(20):6613–6625.

Goldhirsch, A., Glick, J. H., Gelber, R. D., and Senn, H. J. (1998). Meeting highlights: International Consensus Panel on the Treatment of Primary Breast Cancer. *J. Natl. Cancer Inst.* 90(21):1601–1608.

Golub, T. R., Slonim, D. K., Tamayo, P., Huard, C., Gaasenbeek, M., Mesirov, J. P., Coller, H., Loh, M. L., Downing, J. R., Caligiuri, M. A., Bloomfield, C. D., and Lander, E. S. (1999). Molecular classification of cancer: class discovery and class prediction by gene expression monitoring. *Science* 286(5439):531–537.

Gordon, G. J., Jensen, R. V., Hsiao, L. L., Gullans, S. R., Blumenstock, J. E., Ramaswamy, S., Richards, W. G., Sugarbaker, D. J., and Bueno, R. (2002). Translation of microarray data into clinically relevant cancer diagnostic tests using gene expression ratios in lung cancer and mesothelioma. *Cancer Res.* 62(17):4963–4967.

Gordon, G. J., Rockwell, G. N., Godfrey, P. A., Jensen, R. V., Glickman, J. N., Yeap, B. Y., Richards, W. G., Sugarbaker, D. J., and Bueno, R. (2005). Validation of genomics-based prognostic tests in malignant pleural mesothelioma. *Clin. Cancer Res.* 11(12):4406–4414.

Goryachev, A. B., Macgregor, P. F., and Edwards, A. M. (2001). Unfolding of microarray data. *J. Computat. Biol.* 8:443–461.

Greiner, T. C. (2004). mRNA microarray analysis in lymphoma and leukemia. *Cancer Treat Res* 121:1–12 (review).

Grotzer, M. A., Patti, R., Geoerger, B., Eggert, A., Chou, T. T., and Phillips, P. C. (2000). Biological stability of RNA isolated from RNAlater-treated brain tumor and neuroblastoma xenografts. *Med Pediatr Oncol.* 34(6):438–442.

Gruvberger, S., Ringner, M., Chen, Y., Panavally, S., Saal, L. H., Borg, A., Ferno, M., Peterson, C., and Meltzer, P. S. (2001). Estrogen receptor status in breast cancer is associated with remarkably distinct gene expression patterns. *Cancer Res.* 61(16):5979–5984.

Gruvberger, S. K., Ringner, M., Eden, P., Borg, A., Ferno, M., Peterson, C., and Meltzer, P. S. (2002). Expression profiling to predict outcome in breast cancer: the influence of sample selection. *Breast Cancer Res.* 5(1):23–26.

Gruvberger, S. K., Ringner, M., Eden, P., Borg, A., Ferno, M., Peterson, C., and Meltzer, P. S. (2003). Expression profiling to predict outcome in breast cancer: the influence of sample selection. *Breast Cancer Res.* 5(1):23–26.

Gruvberger-Saal, S. K., Eden, P., Ringner, M., Baldetorp, B., Chebil, G., Borg, A., Ferno, M., Peterson, C., and Meltzer, P. S. (2004). Predicting continuous values of prognostic markers in breast cancer from microarray gene expression profiles. *Mol. Cancer Ther.* 3(2):161–168.

Guigo, R., Knudsen, S., Drake, N., and Smith. T. (1992). Prediction of gene structure. *J. Mol. Biol.* 226:141–157.

Gutierrez, N. C., Lopez-Perez, R., Hernandez, J. M., Isidro, I., Gonzalez, B., Delgado, M., Ferminan, E., Garcia, J. L., Vazquez, L., Gonzalez, M., and San Miguel, J. F. (2005). Gene expression profile reveals deregulation of genes with relevant functions in the different subclasses of acute myeloid leukemia. *Leukemia* 19(3):402–409.

Haferlach, T., Kohlmann, A., Kern, W., Hiddemann, W., Schnittger, S., and Schoch, C. (2003). Gene expression profiling as a tool for the diagnosis of acute leukemias. *Semin. Hematol.* 40(4):281–295 (review).

Hastie, T., Tibshirani, R., Eisen, M. B., Alizadeh, A., Levy, R., Staudt, L., Chan, W. C., Botstein, D., and Brown, P. (2000). Gene shaving as a method for identifying distinct sets of genes with similar expression patterns. *Genome Biol.* 1:1–21.

Hastie, T., Tibshirani, R., Botstein, D., and Brown, P. (2001). Supervised harvesting of expression trees. *Genome Biol.* 2:RESEARCH0003, 1–12.

Haviv, I., and Campbell, I. G. (2002). DNA microarrays for assessing ovarian cancer gene expression. *Mol. Cell. Endocrinol.* 191(1):121–126 (review).

Hayashi, Y. (2003). Gene expression profiling in childhood acute leukemia: progress and perspectives. *Int. J. Hematol.* 78(5):414–420 (review).

Hedenfalk, I., Duggan, D., Chen, Y., Radmacher, M., Bittner, M., Simon, R., Meltzer, P., Gusterson, B., Esteller, M., Kallioniemi, O. P., Wilfond, B., Borg, A., and Trent, J. (2001). Gene expression profiles in hereditary breast cancer. *N. Engl. J. Med.* 244:539–548.

Hedges, L. V., and Olkin, I. (1985). *Statistical Methods For Meta-Analysis*. New York: Academic Press.

Heighway, J., Knapp, T., Boyce, L., Brennand, S., Field, J. K., Betticher, D. C., Ratschiller, D., Gugger, M., Donovan, M., Lasek, A., and Rickert, P. (2002). Expression profiling of primary non-small cell lung cancer for target identification. *Oncogene* 21(50):7749–7763.

Henshall, S. M., Afar, D. E., Hiller, J., Horvath, L. G., Quinn, D. I., Rasiah, K. K., Gish, K., Willhite, D., Kench, J. G., Gardiner-Garden, M., Stricker, P. D., Scher, H. I., Grygiel, J. J., Agus, D. B., Mack, D. H., and Sutherland, R. L. (2003). Survival analysis of genome-wide gene expression profiles of prostate cancers identifies new prognostic targets of disease relapse. *Cancer Res.* 63(14):4196–4203.

Herrero, J., Valencia, A., and Dopazo, J. (2001). A hierarchical unsupervised growing neural network for clustering gene expression patterns. *Bioinformatics* 17:126–136.

Hibbs, K., Skubitz, K. M., Pambuccian, S. E., Casey, R. C., Burleson, K. M., Oegema, T. R., Thiele, J. J., Grindle, S. M., Bliss, R. L., and Skubitz, A. P. (2004). Differential gene expression in ovarian carcinoma: identification of potential biomarkers. *Am. J. Pathol.* 165(2):397–414.

Hoang, C. D., D'Cunha, J., Tawfic, S. H., Gruessner, A. C., Kratzke, R. A., and Maddaus, M. A. (2004). Expression profiling of non-small cell lung carcinoma identifies metastatic genotypes based on lymph node tumor burden. *J. Thorac. Cardiovasc. Surg.* 127(5):1332–1341.

Holleman, A., Cheok, M. H., den Boer, M. L., Yang, W., Veerman, A. J., Kazemier, K. M., Pei, D., Cheng, C., Pui, C. H., Relling, M. V., Janka-Schaub, G. E., Pieters, R., and Evans, W. E. (2004). Gene-expression patterns in drug-resistant acute lymphoblastic leukemia cells and response to treatment. *N. Engl. J. Med.* 351(6):533–542.

Holter, N. S., Mitra, M., Maritan, A., Cieplak, M., Banavar, J. R., and Fedoroff, N.V. (2000). Fundamental patterns underlying gene expression profiles: simplicity from complexity. *Proc. Natl. Acad. Sci. USA* 97:8409–8414.

Hough, C. D., Cho, K. R., Zonderman, A. B., Schwartz, D. R., and Morin, P. J. (2001). Coordinately up-regulated genes in ovarian cancer. *Cancer Res.* 61(10):3869–3876.

Houldsworth, J., Olshen, A. B., Cattoretti, G., Donnelly, G. B., Teruya-Feldstein, J., Qin, J., Palanisamy, N., Shen, Y., Dyomina, K., Petlakh, M., Pan, Q., Zelenetz, A. D., Dalla-Favera, R., and Chaganti, R. S. (2004). Relationship between REL amplification, REL function, and clinical and biologic features in diffuse large B-cell lymphomas. *Blood* 103(5):1862–1868.

Huang, E., Cheng, S. H., Dressman, H., Pittman, J., Tsou, M. H., Horng, C. F., Bild, A., Iversen, E. S., Liao, M., Chen, C. M., West, M., Nevins, J. R., and Huang, A. T. (2003). Gene expression predictors of breast cancer outcomes. *Lancet* 361(9369):1590–1596.

Huang, H., Colella, S., Kurrer, M., Yonekawa, Y., Kleihues, P., and Ohgaki, H. (2000). Gene expression profiling of low-grade diffuse astrocytomas by cDNA arrays. *Cancer Res.* 60(24):6868–6874.

Huber, W., Von Heydebreck, A., Sultmann, H., Poustka, A., and Vingron, M. (2002). Variance stabilization applied to microarray data calibration and to the quantification of differential expression. *Bioinformatics* 18 (Suppl 1):S96–S104.

Hughes, T. R., Marton, M. J., Jones, A. R., Roberts, C. J., and Stoughton, R., et. al. (2000). Functional discovery via a compendium of expression profiles. *Cell* 102:109–126.

Hughes, T. R., Mao, M., Jones, A. R., Burchard, J., Marton, M. J., Shannon, K. W., Lefkowitz, S. M., Ziman, M., Schelter, J. M., Meyer, M. R., Kobayashi, S., Davis, C., Dai, H., He, Y. D., Stephaniants, S. B., Cavet, G., Walker, W. L., West, A., Coffey, E., Shoemaker, D. D., Stoughton, R., Blanchard, A. P., Friend, S. H., and Linsley, P. S. (2001). Expression profiling using microarrays fabricated by an ink-jet oligonucleotide synthesizer. *Nature Biotechnol.* 19:342–347.

Huppi, K., and Chandramouli, GV. (2004). Molecular profiling of prostate cancer. *Curr. Urol. Rep.* 5(1):45–51 (review).

Hyvärinen, A. (1999). Fast and robust fixed-point algorithms for independent component analysis. *IEEE Trans. Neural Networks* 10(3):626–634.

Ideker, T., Thorsson, V., Siegel, A. F., and Hood, L. (2000). Testing for differentially-expressed genes by maximum-likelihood analysis of microarray data. *J. Computat. Biol.* 7:805–817.

Irizarry, R. A., Bolstad, B. M., Collin, F., Cope, L. M., Hobbs, B., Speed, T. P. (2003a). Summaries of Affymetrix GeneChip probe level data. *Nucleic Acids Res.* 31(4):e15.

Irizarry, R. A., Hobbs, B., Collin, F., Beazer-Barclay, Y. D., Antonellis, K. J., Scherf, U., and Speed, T. P. (2003b). Exploration, normalization, and summaries of high density oligonucleotide array probe level data. *Biostatistics* 4(2):249–264.

Ismail, R. S., Baldwin, R. L., Fang, J., Browning, D., Karlan, B. Y., Gasson, J. C., and Chang, D. D. (2000). Differential gene expression between normal and tumor-derived ovarian epithelial cells. *Cancer Res.* 60(23):6744–6749.

Jensen, L. J., and Knudsen, S. (2000). Automatic discovery of regulatory patterns in promoter regions based on whole cell expression data and functional annotation. *Bioinformatics* 16:326–333.

Jensen, L. J., Gupta, R., Blom, N., Devos, D., Tamames, J., Kesmir, C., Nielsen, H., Stærfeldt, H. H., Rapacki, K., Workman, C., Andersen, C. A. F., Knudsen, S., Krogh, A., Valencia, A., and Brunak., S. (2002). Ab initio prediction of human orphan protein function from post-translational modifications and localization features. *J. Mol. Biol.* 319:1257–1265.

Jiang, H., Deng, Y., Chen, H. S., Tao, L., Sha, Q., Chen, J., Tsai, C. J., and Zhang, S. (2004). Joint analysis of two microarray gene-expression data sets to select lung adenocarcinoma marker genes. *BMC Bioinformatics* 5(1):81.

Jones, A. M., Douglas, E. J., Halford, S. E., Fiegler, H., Gorman, P. A., Roylance, R. R., Carter, N. P., and Tomlinson, I. P. (2005). Array-CGH analysis of microsatellite-stable, near-diploid bowel cancers and comparison with other types of colorectal carcinoma. *Oncogene* 24(1):118–129.

Kaneta, Y., Kagami, Y., Katagiri, T., Tsunoda, T., Jin-nai, I., Taguchi, H., Hirai, H., Ohnishi, K., Ueda, T., Emi, N., Tomida, A., Tsuruo, T., Nakamura, Y., and Ohno, R. (2002). Prediction of sensitivity to STI571 among chronic myeloid leukemia patients by genome-wide cDNA microarray analysis. *Jpn. J. Cancer Res.* 93(8):849–856.

Kerr, M. K., and Churchill, G. A. (2001a). Statistical design and the analysis of gene expression microarray data. *Genet. Res.* 77:123–128 (review).

Kerr, M. K., and Churchill, G. A. (2001b). Bootstrapping cluster analysis: assessing the reliability of conclusions from microarray experiments. *Proc. Natl. Acad. Sci. USA* 98:8961–8965.

Kerr, M. K., Martin, M., and Churchill, G. A. (2000). Analysis of variance for gene expression microarray data. *J. Computat. Biol.* 7:819–837.

Khan, J., Wei, J. S., Ringner, M., Saal, L. H., Ladanyi, M., Westermann, F., Berthold, F., Schwab, M., Antonescu, C. R., Peterson, C., and Meltzer, P. S. (2001). Classification and diagnostic prediction of cancers using gene expression profiling and artificial neural networks. *Nature Genet.* 7:673–679.

Khatua, S., Peterson, K. M., Brown, K. M., Lawlor, C., Santi, M. R., LaFleur, B., Dressman, D., Stephan, D. A., and MacDonald, T. J. (2003). Overexpression of the EGFR/FKBP12/HIF-2alpha pathway identified in childhood astrocytomas by angiogenesis gene profiling. *Cancer Res.* 63(8):1865–1870.

Kikuchi, T., Daigo, Y., Katagiri, T., Tsunoda, T., Okada, K., Kakiuchi, S., Zembutsu, H., Furukawa, Y., Kawamura, M., Kobayashi, K., Imai, K., and Nakamura, Y. (2003). Expression profiles of non-small cell lung cancers on cDNA microarrays: identification of genes for prediction of lymph-node metastasis and sensitivity to anti-cancer drugs. *Oncogene* 22(14):2192–2205.

Kim, C. J., Reintgen, D. S., and Yeatman, T. J. (2002). The promise of microarray technology in melanoma care. *Cancer Control* 9(1):49–53(review).

Kim, S., Dougherty, E. R., Shmulevich, L., Hess, K. R., Hamilton, S. R., Trent, J. M., Fuller, G. N., and Zhang, W. (2002). Identification of combination gene sets for glioma classification. *Mol. Cancer Ther.* 1(13):1229-1236.

Knudsen, S. (1999). Promoter2.0: for the recognition of PolII promoter sequences. *Bioinformatics* 15:356–361. (Available as a web server at http://www.cbs.dtu.dk/services/Promoter/.)

Kohlmann, A., Schoch, C., Schnittger, S., Dugas, M., Hiddemann, W., Kern, W., and Haferlach, T. (2003). Molecular characterization of acute leukemias by use of microarray technology. *Genes Chromosomes Cancer* 37(4):396–405.

Kohlmann, A., Schoch, C., Schnittger, S., Dugas, M., Hiddemann, W., Kern, W., and Haferlach, T. (2004). Pediatric acute lymphoblastic leukemia (ALL) gene expression signatures classify an independent cohort of adult ALL patients. *Leukemia* 18(1):63–71.

Kohonen, T. (1995). *Self-Organizing Maps.* Berlin: Springer.

Korenberg, M. J. (2003). Gene expression monitoring accurately predicts medulloblastoma positive and negative clinical outcomes. *FEBS Lett.* 533(1-3):110–114.

Korshunov, A., Neben, K., Wrobel, G., Tews, B., Benner, A., Hahn, M., Golanov, A., and Lichter, P. (2003). Gene expression patterns in ependymomas correlate with tumor location, grade, and patient age. *Am. J. Pathol.* 163(5):1721–1727.

Kristiansen, G., Pilarsky, C., Wissmann, C., Kaiser, S., Bruemmendorf, T., Roepcke, S., Dahl, E., Hinzmann, B., Specht, T., Pervan, J., Stephan, C., Loening, S., Dietel, M., and Rosenthal, A. (2005). Expression profiling of microdissected matched prostate cancer samples reveals CD166/MEMD and CD24 as new prognostic markers for patient survival. *J. Pathol.* 205(3):359–376.

Krogh, A. (1997). Two methods for improving performance of an HMM and their application for gene finding. *Proceedings Fifth International Conference on Intelligent Systems for Molecular Biology (ISMB)*. Menlo Park, CA: AAAI Press, pp. 179–186.

Kruhoffer, M., Jensen, J. L., Laiho, P., Dyrskjot, L., Salovaara, R., Arango, D., Birkenkamp-Demtroder, K., Sorensen, F. B., Christensen, L. L., Buhl, L., Mecklin, J. P., Jarvinen, H., Thykjaer, T., Wikman, F. P., Bech-Knudsen, F., Juhola, M., Nupponen, N. N., Laurberg, S., Andersen, C. L., Aaltonen, L. A., and Orntoft, T. F. (2005). Gene expression signatures for colorectal cancer microsatellite status and HNPCC. *Br. J. Cancer* 92(12):2240–2248.

Lacayo, N. J., Meshinchi, S., Kinnunen, P., Yu, R., Wang, Y., Stuber, C. M., Douglas, L., Wahab, R., Becton, D. L., Weinstein, H., Chang, M. N., Willman, C. L., Radich, J. P., Tibshirani, R., Ravindranath, Y., Sikic, B., and Dahl, G. V. (2004). Gene expression profiles at diagnosis in de novo childhood AML patients identify FLT3 mutations with good clinical outcomes. *Blood* 104(9):2645–2654.

Lancaster, J. M., Dressman, H. K., Whitaker, R. S., Havrilesky, L., Gray, J., Marks, J. R., Nevins, J. R., and Berchuck, A. (2004). Gene expression patterns that characterize advanced stage serous ovarian cancers. *J. Soc. Gynecol. Investig.* 11(1):51–59.

Lapointe, J., Li, C., Higgins, J. P., van de Rijn, M., Bair, E., Montgomery, K., Ferrari, M., Egevad, L., Rayford, W., Bergerheim, U., Ekman, P., DeMarzo, A. M., Tibshirani, R., Botstein, D., Brown, P. O., Brooks, J. D., and Pollack, J. R. (2004). Gene expression profiling identifies clinically relevant subtypes of prostate cancer. *Proc. Natl. Acad. Sci. USA* 101(3):811–816.

Lash, A. E., Tolstoshev, C. M., Wagner, L., Schuler, G. D., Strausberg, R. L., Riggins, G. J., and Altschul, S. F. (2000). SAGEmap: a public gene expression resource. *Genome Res.* 10:1051–1060.

Latil, A., Bieche, I., Chene, L., Laurendeau, I., Berthon, P., Cussenot, O., and Vidaud, M. (2003). Gene expression profiling in clinically localized prostate cancer: a four-gene expression model predicts clinical behavior. *Clin. Cancer Res.* 9(15):5477–5485.

LaTulippe, E., Satagopan, J., Smith, A., Scher, H., Scardino, P., Reuter, V., and Gerald, W. L. (2002). Comprehensive gene expression analysis of prostate cancer reveals distinct transcriptional programs associated with metastatic disease. *Cancer Res.* 62(15):4499–4506.

Lawrence, C. E., Altschul, S. F., Boguski, M. S., Liu, J. S., Neuwald, A. F., and Wootton, J. C. (1993). Detecting subtle sequence signals: a Gibbs sampling strategy for multiple alignment. *Science* 262:208–214.

Lazaridis, E. N., Sinibaldi, D., Bloom, G., Mane, S., and Jove, R. (2002). A simple method to improve probe set estimates from oligonucleotide arrays. *Math. Biosci.* 176(1):53–58.

Lee, B. C., Cha, K., Avraham, S., and Avraham, H. K. (2004). Microarray analysis of differentially expressed genes associated with human ovarian cancer. *Int. J. Oncol.* 24(4):847–851.

Lee, M. L., and Whitmore, G. A. (2002). Power and sample size for DNA microarray studies. *Stat. Med.* 21(23):3543–3570.

Lee, S. I., and Batzoglou, S. (2003). Application of independent component analysis to microarrays. *Genome Biol.* 4(11):R76. Epub 2003 Oct 24.

Lemon, W. J., Palatini, J. T., Krahe, R., and Wright, F. A. (2002). Theoretical and experimental comparisons of gene expression indexes for oligonucleotide arrays. *Bioinformatics* 18(11):1470–1476.

Lemon, W. J., Liyanarachchi, S., and You, M. (2003). A high performance test of differential gene expression for oligonucleotide arrays. *Genome Biol.* 4(10):R67.

Li, C., and Wong, W. H. (2001a). Model-based analysis of oligonucleotide arrays: expression index computation and outlier detection. *Proc. Natl. Acad. Sci. USA* 98:31–36. (Software available at http://www.dchip.org.)

Li, C., and Wong, W. H. (2001b). Model-based analysis of oligonucleotide arrays: Model validation, design issues and standard error application. *Genome Biol.* 2:1–11. (Software available at http://www.dchip.org.)

Li, F., and Stormo, G. D. (2001). Selection of optimal DNA oligos for gene expression arrays. *Bioinformatics* 17:1067–1076.

Liebermeister, W. (2002). Linear modes of gene expression determined by independent component analysis. *Bioinformatics* 18(1):51–60.

Lipshutz, R. J., Fodor, S. P. A., Gingeras, T. R., and Lockhart, D. J. (1999). High density synthetic oligonucleotide arrays. *Nature Genet. Chipping Forecast* 21:20–24.

Liu, X., Brutlag, D. L., and Liu, J. S. (2001). BioProspector: discovering conserved DNA motifs in upstream regulatory regions of co-expressed genes. *Pacific Symposium on Biocomputing* 6:127–138. (Available online at http://psb.stanford.edu.)

Ljubimova, J. Y., Khazenzon, N. M., Chen, Z., Neyman, Y. I., Turner, L., Riedinger, M. S., and Black, K. L. (2001). Gene expression abnormalities in human glial tumors identified by gene array. *Int. J. Oncol.* 18(2):287–295.

Lockhart, D. J., Dong, H., Byrne, M. C., Follettie, M. T., Gallo, M. V., Chee, M. S., Mittmann, M., Wang C., Kobayashi, M., Horton, H., and Brown, E. L. (1996). Expression monitoring by hybridization to high-density oligonucleotide arrays. *Nature Biotechnol.* 14:1675–1680.

Lonnstedt, I., and Speed, T. (2002). Replicated microarray data. *Statistica Sinica* 12:31–46.

Lossos, I. S., Jones, C. D., Warnke, R., Natkunam, Y., Kaizer, H., Zehnder, J. L., Tibshirani, R., and Levy, R. (2001). Expression of a single gene, BCL-6, strongly predicts survival in patients with diffuse large B-cell lymphoma. *Blood* 98(4):945–951.

Lu, J., Getz, G., Miska, E. A., Alvarez-Saavedra, E., Lamb, J., Peck, D., Sweet-Cordero, A., Ebert, B. L., Mak, R. H., Ferrando, A. A., Downing, J. R., Jacks, T., Horvitz, H. R., and Golub, T. R. (2005). MicroRNA expression profiles classify human cancers. *Nature* 435(7043):834–838.

Lu, P., Nakorchevskiy, A., and Marcotte, E. M. (2003). Expression deconvolution: a reinterpretation of DNA microarray data reveals dynamic changes in cell populations. *Proc. Natl. Acad. Sci. USA* 100(18):10370–10375.

Luo, J. H., Yu, Y. P., Cieply, K., Lin, F., Deflavia, P., Dhir, R., Finkelstein, S., Michalopoulos, G., and Becich, M. (2002). Gene expression analysis of prostate cancers. *Mol. Carcinog.* 33(1):25–35.

Magee, J. A., Araki, T., Patil, S., Ehrig, T., True, L., Humphrey, P. A., Catalona, W. J., Watson, M. A., and Milbrandt, J. (2001). Expression profiling reveals hepsin overexpression in prostate cancer. *Cancer Res.* 61(15):5692–5696.

Man, M. Z., Wang, X., and Wang, Y. (2000). POWER_SAGE: comparing statistical tests for SAGE experiments. *Bioinformatics* 16:953–959.

Margulies, E. H., and Innis, J. W. (2000). eSAGE: managing and analysing data generated with serial analysis of gene expression (SAGE). *Bioinformatics* 16:650–651.

Marie, R., Jensenius, H., Thaysen, J. Christensen, C. B., and Boisen, A. (2002). Adsorption kinetics and mechanical properties of thiol-modied DNA-oligos on gold investigated by microcantilever sensors. *Ultramicroscopy* 91:29–36.

Martin, K. J., Kritzman, B. M., Price, L. M., Koh, B., Kwan, C. P., Zhang, X., Mackay, A., O'Hare, M. J., Kaelin, C. M., Mutter, G. L., Pardee, A. B., and Sager, R. (2000). Linking gene expression patterns to therapeutic groups in breast cancer. *Cancer Res.* 60(8):2232–2238.

Martoglio, A. M., Tom, B. D., Starkey, M., Corps, A. N., Charnock-Jones, D. S., and Smith, S. K. (2000). Changes in tumorigenesis- and angiogenesis-related gene transcript abundance profiles in ovarian cancer detected by tailored high density cDNA arrays. *Mol. Med.* 6(9):750–765.

Martoglio, A. M., Miskin, J. W., Smith, S. K., and MacKay, D. J. A decomposition model to track gene expression signatures: preview on observer-independent classification of ovarian cancer. (2002). *Bioinformatics* 18(12):1617–1624.

Matei, D., Graeber, T. G., Baldwin, R. L., Karlan, B. Y., Rao, J., and Chang, D. D. (2002). Gene expression in epithelial ovarian carcinoma. *Oncogene* 21(41):6289–6298.

Matthews, B. W. (1975). Comparison of the predicted and observed secondary structure of T4 phage lysozyme. *Biochim. Biophys. Acta* 405:442–451.

Matz, M., Usman, N., Shagin, D., Bogdanova, E., and Lukyanov, S. (1997). Ordered differential display: a simple method for systematic comparison of gene expression profiles. *Nucleic Acids Res.* 25:2541–2542.

McDonald, S. L., Edington, H. D., Kirkwood, J. M., and Becker, D. (2004). Expression analysis of genes identified by molecular profiling of VGP melanomas and MGP melanoma-positive lymph nodes. *Cancer Biol. Ther.* 3(1):110–120.

McKendry, R., Zhang, J., Arntz, Y., Strunz, T., Hegner, M., Lang, H. P., Baller, M. K., Certa, U., Meyer, E., Guntherodt, H. J., and Gerber, C. (2002). Multiple label-free biodetection and quantitative DNA-binding assays on a nanomechanical cantilever array. *Proc. Natl. Acad. Sci. USA* 99:9783–9788.

Mecham, B. H., Klus, G. T., Strovel, J., Augustus, M., Byrne, D., Bozso, P., Wetmore, D. Z., Mariani, T. J., Kohane, I. S., and Szallasi, Z. (2004). Sequence-matched probes produce increased cross-platform consistency and more reproducible biological results in microarray-based gene expression measurements. *Nucleic Acids Res.* 32(9):e74.

Michaels, G. S., Carr, D. B., Askenazi, M., Fuhrman, S., Wen, X., and Somogyi, R. (1998). Cluster analysis and data visualization of large-scale gene expression data. *Pacific Symposium on Biocomputing* 3:42–53. (Available online at http://psb.stanford.edu.)

Miller, L. D., Smeds, J., George, J., Vega, V. B., Vergara, L., Ploner, A., Pawitan, Y., Hall, P., Klaar, S., Liu, E. T., and Bergh, J. (2005). An expression signature for p53 status in human breast cancer predicts mutation status, transcriptional effects, and patient survival. *Proc. Natl. Acad. Sci. USA* 102(38):13550–13555.

Minn, A. J., Gupta, G. P., Siegel, P. M., Bos, P. D., Shu, W., Giri, D. D., Viale, A., Olshen, A. B., Gerald, W. L., and Massague, J. (2005). Genes that mediate breast cancer metastasis to lung. *Nature* 436(7050):518–524.

Mischel, P. S., Shai, R., Shi, T., Horvath, S., Lu, K. V., Choe, G., Seligson, D., Kremen, T. J., Palotie, A., Liau, L. M., Cloughesy, T. F., and Nelson, S. F. (2003). Identification of molecular subtypes of glioblastoma by gene expression profiling. *Oncogene* 22(15):2361–2373.

Miura, K., Bowman, E. D., Simon, R., Peng, A. C., Robles, A. I., Jones, R. T., Katagiri, T., He, P., Mizukami, H., Charboneau, L., Kikuchi, T., Liotta, L. A., Nakamura, Y., and Harris, C. C. (2002). Laser capture microdissection and microarray expression analysis of lung adenocarcinoma reveals tobacco smoking- and prognosis-related molecular profiles. *Cancer Res.* 62(11):3244–3250.

Modlich, O., Prisack, H. B., Pitschke, G., Ramp, U., Ackermann, R., Bojar, H., Vogeli, T. A., and Grimm, M. O. (2004). Identifying superficial, muscle-invasive, and metastasizing transitional cell carcinoma of the bladder: use of cDNA array analysis of gene expression profiles. *Clin. Cancer Res.* 10(10):3410–3421.

Modlich, O., Prisack, H. B., Munnes, M., Audretsch, W., and Bojar, H. (2005). Predictors of primary breast cancers responsiveness to preoperative epirubicin/cyclophosphamide-based chemotherapy: translation of microarray data into clinically useful predictive signatures. *J. Transl. Med.* 3:32.

Montgomery, D. C., and Runger, G. C. (1999). *Applied Statistics and Probability for Engineers.* Hoboken, NJ: Wiley.

Monti, S., Savage, K. J., Kutok, J. L., Feuerhake, F., Kurtin, P., Mihm, M., Wu, B., Pasqualucci, L., Neuberg, D., Aguiar, R. C., Dal, C. I., Ladd, C., Pinkus, G. S., Salles, G., Harris, N. L., Dalla-Favera, R., Habermann, T. M., Aster, J. C., Golub, T. R., and Shipp, M. A. (2005). Molecular profiling of diffuse large B-cell lymphoma identifies robust subtypes including one characterized by host inflammatory response. *Blood* 105(5):1851–1861.

Moos, P. J., Raetz, E. A., Carlson, M. A., Szabo, A., Smith, F. E., Willman, C., Wei, Q., Hunger, S. P., and Carroll, W. L. (2002). Identification of gene expression profiles that segregate patients with childhood leukemia. *Clin. Cancer. Res.* 8(10):3118–3130.

Moreau, Y., Aerts, S., De Moor, B., De Strooper, B., and Dabrowski, M. (2003). Comparison and meta-analysis of microarray data: from the bench to the computer desk. *Trends Genet.*19(10):570–577 (review).

Muller, M. C., Hordt, T., Paschka, P., Merx, K., La, R. o., Hehlmann, R., and Hochhaus, A. (2004). Standardization of preanalytical factors for minimal residual disease analysis in chronic myelogenous leukemia. *Acta Haematol.* 112(1-2):30–33. Review.

Mutter, G. L., Zahrieh, D., Liu, C., Neuberg, D., Finkelstein, D., Baker, H. E., and Warrington, J. A. (2004). Comparison of frozen and RNALater solid tissue storage methods for use in RNA expression microarrays. *BMC Genomics* 5(1):88.

Nagahata, T., Onda, M., Emi, M., Nagai, H., Tsumagari, K., Fujimoto, T., Hirano, A., Sato, T., Nishikawa, K., Akiyama, F., Sakamoto, G., Kasumi, F., Miki, Y., Tanaka, T., and Tsunoda, T. (2004). Expression profiling to predict postoperative prognosis for estrogen receptor-negative breast cancers by analysis of 25,344 genes on a cDNA microarray. *Cancer Sci.* 95(3):218–225.

Nawrocki, S., Skacel, T., and Brodowicz, T. (2003). From microarrays to new therapeutic approaches in bladder cancer. *Pharmacogenomics* 4(2):179–189 (review).

Neben, K., Korshunov, A., Benner, A., Wrobel, G., Hahn, M., Kokocinski, F., Golanov, A., Joos, S., and Lichter, P. (2004). Microarray-based screening for molecular markers in medulloblastoma revealed STK15 as independent predictor for survival. *Cancer Res.* 64(9):3103–3111.

Neuwald, A. F., Liu, J. S., and Lawrence, C. E. (1995). Gibbs motif sampling: detection of bacterial outer membrane protein repeats. *Protein Sci.* 4:1618–1632.

Newton, M. A., Kendziorski, C. M., Richmond, C. S., Blattner, F. R., and Tsui, K. W. (2001). On differential variability of expression ratios: improving statistical inference about gene expression changes from microarray data. *J. Comput. Biol.* 8:37–52.

Nguyen, D. V., and Rocke, D. M. (2002a). Multi-class cancer classification via partial least squares with gene expression profiles. *Bioinformatics* 18(9):1216–1226.

Nguyen, D. V., and Rocke, D. M. (2002b). Partial least squares proportional hazard regression for application to DNA microarray survival data. *Bioinformatics* 18(12):1625–1632.

Nguyen, D. V., and Rocke, D. M. (2002c). Tumor classification by partial least squares using microarray gene expression data. *Bioinformatics* 18(1):39–50.

Nielsen, H. B., and Knudsen, S. (2002). Avoiding cross hybridization by choosing nonredundant targets on cDNA arrays. *Bioinformatics* 18:321–322.

Nielsen, H. B., Wernersson, R., and Knudsen, S. (2003). Design of oligonucleotides for microarrays and perspectives for design of multi-transcriptome arrays. *Nucleic Acids Res.* 31:3491–3496.

Nishizuka, S., Chen, S. T., Gwadry, F. G., Alexander, J., Major, S. M., Scherf, U., Reinhold, W. C., Waltham, M., Charboneau, L., Young, L., Bussey, K. J., Kim, S., Lababidi, S., Lee, J. K., Pittaluga, S., Scudiero, D. A., Sausville, E. A., Munson, P. J., Petricoin, E. F., Liotta, L. A., Hewitt, S. M., Raffeld, M., and Weinstein, J. N. (2003). Diagnostic markers that distinguish colon and ovarian adenocarcinomas: identification by genomic, proteomic, and tissue array profiling. *Cancer Res.* 63(17):5243–5250.

Nocito, A., Bubendorf, L., Maria, T. I., Suess, K., Wagner, U., Forster, T., Kononen, J., Fijan, A., Bruderer, J., Schmid, U., Ackermann, D., Maurer, R., Alund, G., Knonagel, H., Rist, M., Anabitarte, M., Hering, F., Hardmeier, T., Schoenenberger, A. J., Flury, R., Jager, P., Luc Fehr, J., Schraml, P., Moch, H., Mihatsch, M. J., Gasser, T., and Sauter, G. (2001). Microarrays of bladder cancer tissue are highly representative of proliferation index and histological grade. *J. Pathol.* 194(3):349–357.

Notterman, D. A., Alon, U., Sierk, A. J., and Levine, A. J. (2001). Transcriptional gene expression profiles of colorectal adenoma, adenocarcinoma, and normal tissue examined by oligonucleotide arrays. *Cancer Res.* 61(7):3124–3130.

Nutt, C. L., Mani, D. R., Betensky, R. A., Tamayo, P., Cairncross, J. G., Ladd, C., Pohl, U., Hartmann, C., McLaughlin, M. E., Batchelor, T. T., Black, P. M., von Deimling, A., Pomeroy, S. L., Golub, T. R., and Louis, D. N. (2003). Gene expression-based classification of malignant gliomas correlates better with survival than histological classification. *Cancer Res.* 63(7):1602–1607.

Nuwaysir, E. F., Huang, W., Albert, T. J., Singh, J., Nuwaysir, K., Pitas, A., Richmond, T., Gorski, T., Berg, J. P., Ballin, J., McCormick, M., Norton, J., Pollock, T., Sumwalt, T., Butcher, L., Porter, D., Molla, M., Hall, C., Blattner, F., Sussman, M. R., Wallace, R. L., Cerrina, F., and Green, R. D. (2002). Gene expression analysis using oligonucleotide arrays produced by maskless photolithography. *Genome Res.* 12:1749–1755.

Ohashi, Y., Creek, K. E., Pirisi, L., Kalus, R., and Young, S. R. (2004). RNA degradation in human breast tissue after surgical removal: a time-course study. *Exp. Mol. Pathol.* 77(2):98–103.

Onda, M., Emi, M., Nagai, H., Nagahata, T., Tsumagari, K., Fujimoto, T., Akiyama, F., Sakamoto, G., Makita, M., Kasumi, F., Miki, Y., Tanaka, T., Tsunoda, T., and Nakamura, Y. (2004). Gene expression patterns as marker for 5-year postoperative prognosis of primary breast cancers. *J. Cancer Res. Clin. Oncol.* 130(9):537–545.

Oshima, Y., Ueda, M., Yamashita, Y., Choi, Y. L., Ota, J., Ueno, S., Ohki, R., Koinuma, K., Wada, T., Ozawa, K., Fujimura, A., and Mano, H. (2003). DNA microarray analysis of hematopoietic stem cell-like fractions from individuals with the M2 subtype of acute myeloid leukemia. *Leukemia* 17(10):1990–1997.

Ouzounis, C. A., and Valencia, A. (2003). Early bioinformatics: the birth of a discipline—a personal view. *Bioinformatics* 19(17):2176–2190.

Pan, W., Lin, J., and Le, C. (2001). How many replicates of arrays are required to detect gene expression changes in microarray experiments? A mixture model approach. Report 2001-012, Division of Biostatistics, University of Minnesota. (Available at http://www.biostat.umn.edu/cgi-bin/rrs?print+2001.)

Pan, W., Lin, J., and Le, C. (2002). How many replicates of arrays are required to detect gene expression changes in microarray experiments? A mixture model approach. *Genome Biology* 3(5):research0022, 1–10.

Park, P. C., Taylor, M. D., Mainprize, T. G., Becker, L. E., Ho, M., Dura, W. T., Squire, J., and Rutka, J. T. (2003). Transcriptional profiling of medulloblastoma in children. *J. Neurosurg.* 99(3):534–541.

Park, P. J., Pagano, M., and Bonetti, M. (2001). A nonparametric scoring algorithm for identifying informative genes from microarray data. *Pacific Symposium on Biocomputing* 6:52–63. (Manuscript available online at http://psb.stanford.edu.)

Parmigiani, G., Garrett, E. S., Irizarry, R. A., and Zeger, S. (Editors). (2003). *The Analysis of Gene Expression Data*. Berlin: Springer Verlag.

Parmigiani, G., Garrett-Mayer, E. S., Anbazhagan, R., and Gabrielson, E. (2004). A cross-study comparison of gene expression studies for the molecular classification of lung cancer. *Clin. Cancer Res.* 10(9):2922–2927.

Pavey, S., Johansson, P., Packer, L., Taylor, J., Stark, M., Pollock, P. M., Walker, G. J., Boyle, G. M., Harper, U., Cozzi, S. J., Hansen, K., Yudt, L., Schmidt, C., Hersey, P.,

Ellem, K. A., O'Rourke, M. G., Parsons, P. G., Meltzer, P., Ringner, M., and Hayward, N. K. (2004). Microarray expression profiling in melanoma reveals a BRAF mutation signature. *Oncogene* 23(23):4060–4067.

Pavlidis, P., Li, Q., and Noble, W. S. (2003). The effect of replication on gene expression microarray experiments. *Bioinformatics* 19(13):1620–1627.

Pawitan, Y., Bjohle. J., Amler, L., Borg, A.-L., Egyhazi, S., Hall, P., Han, X., Holmberg, L., Huang, F., Klaar, S., Liu, E., Miller, L., Nordgren, H., Ploner, A., Sandelin, K., Shaw, P., Smeds, J., Skoog, L., Wedren, S., and Bergh, J. (2005). Gene expression profiling spares early breast cancer patients from adjuvant therapy: derived and validated in two population-based cohorts *Breast Cancer Res.* 7:R953–R964.

Pearson, W. R. (2001). Training for bioinformatics and computational biology (editorial). *Bioinformatics* 17:761–762.

Perez-Enciso, M., and Tenenhaus, M. (2003). Prediction of clinical outcome with microarray data: a partial least squares discriminant analysis (PLS-DA) approach. *Hum. Genet.* 112(5-6):581–592.

Perou, C. M., Jeffrey, S. S., van de Rijn, M., Rees, C. A., Eisen, M. B., Ross, D. T., Pergamenschikov, A., Williams, C. F., Zhu, S. X., Lee, J. C., Lashkari, D., Shalon, D., Brown, P. O., and Botstein, D. (1999). Distinctive gene expression patterns in human mammary epithelial cells and breast cancers. *Proc. Natl. Acad. Sci.* U S A. 96(16):9212–9217.

Perou, C. M., Sorlie, T., Eisen, M. B., van de Rijn, M., Jeffrey, S. S., Rees, C. A., Pollack, J. R., Ross, D. T., Johnsen, H., Akslen, L. A., Fluge, O., Pergamenschikov, A., Williams, C., Zhu, S. X., Lonning, P. E., Borresen-Dale, A. L., Brown, P. O., and Botstein, D. (2000). Molecular portraits of human breast tumours. *Nature* 406(6797):747–752.

Piper, M. D., Daran-Lapujade, P., Bro, C., Regenberg, B., Knudsen, S., Nielsen, J., and Pronk, J. T. (2002). Reproducibility of oligonucleotide microarray transcriptome analyses. An interlaboratory comparison using chemostat cultures of *Saccharomyces cerevisiae. J. Biol. Chem.* 277(40):37001–37008.

Pittman, J., Huang, E., Dressman, H., Horng, C. F., Cheng, S. H., Tsou, M. H., Chen, C. M., Bild, A., Iversen, E. S., Huang, A. T., Nevins, J. R., and West, M. (2004). Integrated modeling of clinical and gene expression information for personalized prediction of disease outcomes. *Proc. Natl. Acad. Sci. USA* 101(22):8431–8436.

Pochet, N., De Smet, F., Suykens, J. A., and De Moor, B. L. (2004). Systematic benchmarking of microarray data classification: assessing the role of nonlinearity and dimensionality reduction. *Bioinformatics* 20(17):3185–3195.

Pollack, J. R., Sorlie, T., Perou, C. M., Rees, C. A., Jeffrey, S. S., Lonning, P. E., Tibshirani, R., Botstein, D., Borresen-Dale, A. L., and Brown, P. O. (2002). Microarray analysis reveals a major direct role of DNA copy number alteration in the transcriptional program of human breast tumors. *Proc. Natl. Acad. Sci. USA* 99(20):12963–12968.

Pomeroy, S. L., Tamayo, P., Gaasenbeek, M., Sturla, L. M., Angelo, M., McLaughlin, M. E., Kim, J. Y., Goumnerova, L. C., Black, P. M., Lau, C., Allen, J. C., Zagzag, D., Olson, J. M., Curran, T., Wetmore, C., Biegel, J. A., Poggio, T., Mukherjee, S., Rifkin, R., Califano, A., Stolovitzky, G., Louis, D. N., Mesirov, J. P., Lander, E. S., and Golub, T. R. (2002). Prediction of central nervous system embryonal tumour outcome based on gene expression. *Nature* 415(6870):436–442.

Presneau, N., Mes-Masson, A. M., Ge, B., Provencher, D., Hudson, T. J., and Tonin, P. N. (2003). Patterns of expression of chromosome 17 genes in primary cultures of normal ovarian surface epithelia and epithelial ovarian cancer cell lines. *Oncogene* 22(10):1568–1579.

Ramaswamy, S., Tamayo, P., Rifkin, R., Mukherjee, S., Yeang, C. H., Angelo, M., Ladd, C., Reich, M., Latulippe, E., Mesirov, J. P., Poggio, T., Gerald, W., Loda, M., Lander, E. S., and Golub, T. R. (2001). Multiclass cancer diagnosis using tumor gene expression signatures. *Proc. Natl. Acad. Sci. USA* 98(26):15149–15154.

Raychaudhuri, S., Stuart, J. M., and Altman, R. B. (2000). Principal components analysis to summarize microarray experiments: application to sporulation time series. *Pacific Symposium on Biocomputing* 2000:455–466. (Available online at http://psb.stanford.edu.)

Reiner, A., Yekutieli, D., Benjamini, Y. (2003). Identifying differentially expressed genes using false discovery rate controlling procedures. *Bioinformatics* 19(3):368–375.

Rhodes, D. R., Barrette, T. R., Rubin, M. A., Ghosh, D., and Chinnaiyan, A. M. (2002). Meta-analysis of microarrays: interstudy validation of gene expression profiles reveals pathway dysregulation in prostate cancer. *Cancer Res.* 62(15):4427–4433.

Rhodes, D. R., Yu, J., Shanker, K., Deshpande, N., Varambally, R., Ghosh, D., Barrette, T., Pandey, A., and Chinnaiyan, A. M. (2004). Large-scale meta-analysis of cancer microarray data identifies common transcriptional profiles of neoplastic transformation and progression. *Proc. Natl. Acad. Sci. USA* 101(25):9309–9314.

Rickman, D. S., Bobek, M. P., Misek, D. E., Kuick, R., Blaivas, M., Kurnit, D. M., Taylor, J., and Hanash, S. M. (2001). Distinctive molecular profiles of high-grade and low-grade gliomas based on oligonucleotide microarray analysis. *Cancer Res.* 61(18):6885–6891.

Rocke, D. M., and Durbin, B. (2001). A model for measurement error for gene expression arrays. *J. Computat. Biol.* 8(6):557–569.

Roesch, A., Vogt, T., Stolz, W., Dugas, M., Landthaler, M., and Becker, B. (2003). Discrimination between gene expression patterns in the invasive margin and the tumour core of malignant melanomas. *Melanoma Res.* 13(5):503–509.

Roos-van Groningen, M. C., Eikmans, M., Baelde, H. J., de Heer, E., and Bruijn, J. A. (2004). Improvement of extraction and processing of RNA from renal biopsies. *Kidney Int.* 65(1):97–105.

Rosenwald, A., and Staudt, L. M. (2003). Gene expression profiling of diffuse large B-cell lymphoma. *Leuk. Lymphoma* 44 (Suppl 3):S41–S47.

Rosenwald, A., Wright, G., Chan, W. C., Connors, J. M., Campo, E., Fisher, R. I., Gascoyne, R. D., Muller-Hermelink, H. K., Smeland, E. B., Giltnane, J. M., Hurt, E. M., Zhao, H., Averett, L., Yang, L., Wilson, W. H., Jaffe, E. S., Simon, R., Klausner, R. D., Powell, J., Duffey, P. L., Longo, D. L., Greiner, T. C., Weisenburger, D. D., Sanger, W. G., Dave, B. J., Lynch, J. C., Vose, J., Armitage, J. O., Montserrat, E., Lopez-Guillermo, A., Grogan, T. M., Miller, T. P., LeBlanc, M., Ott, G., Kvaloy, S., Delabie, J., Holte, H., Krajci, P., Stokke, T., and Staudt, L. M. (2003a). The use of molecular profiling to predict survival after chemotherapy for diffuse large-B-cell lymphoma. *N. Engl. J. Med.* 346(25):1937–1947.

Rosenwald, A., Wright, G., Leroy, K., Yu, X., Gaulard, P., Gascoyne, R. D., Chan, W. C., Zhao, T., Haioun, C., Greiner, T. C., Weisenburger, D. D., Lynch, J. C., Vose, J.,

Armitage, J. O., Smeland, E. B., Kvaloy, S., Holte, H., Delabie, J., Campo, E., Montserrat, E., Lopez-Guillermo, A., Ott, G., Muller-Hermelink, H. K., Connors, J. M., Braziel, R., Grogan, T. M., Fisher, R. I., Miller, T. P., LeBlanc, M., Chiorazzi, M., Zhao, H., Yang, L., Powell, J., Wilson, W. H., Jaffe, E. S., Simon, R., Klausner, R. D., and Staudt, L. M. (2003b). Molecular diagnosis of primary mediastinal B cell lymphoma identifies a clinically favorable subgroup of diffuse large B cell lymphoma related to Hodgkin lymphoma. *J. Exp. Med.* 198(6):851–862.

Rosenwald, A., Wright, G., Wiestner, A., Chan, W. C., Connors, J. M., Campo, E., Gascoyne, R. D., Grogan, T. M., Muller-Hermelink, H. K., Smeland, E. B., Chiorazzi, M., Giltnane, J. M., Hurt, E. M., Zhao, H., Averett, L., Henrickson, S., Yang, L., Powell, J., Wilson, W. H., Jaffe, E. S., Simon, R., Klausner, R. D., Montserrat, E., Bosch, F., Greiner, T. C., Weisenburger, D. D., Sanger, W. G., Dave, B. J., Lynch, J. C., Vose, J., Armitage, J. O., Fisher, R. I., Miller, T. P., LeBlanc, M., Ott, G., Kvaloy, S., Holte, H., Delabie, J., and Staudt, L. M. (2003c). The proliferation gene expression signature is a quantitative integrator of oncogenic events that predicts survival in mantle cell lymphoma. *Cancer Cell* 3(2):185–197.

Ross, M. E., Zhou, X., Song, G., Shurtleff, S. A., Girtman, K., Williams, W. K., Liu, H. C., Mahfouz, R., Raimondi, S. C., Lenny, N., Patel, A., and Downing, J. R. (2003). Classification of pediatric acute lymphoblastic leukemia by gene expression profiling. *Blood* 102(8):2951–2959. Epub 2003 May 01.

Ross, M. E., Mahfouz, R., Onciu, M., Liu, H. C., Zhou, X., Song, G., Shurtleff, S. A., Pounds, S., Cheng, C., Ma, J., Ribeiro, R. C., Rubnitz, J. E., Girtman, K., Williams, W. K., Raimondi, S. C., Liang, D. C., Shih, L. Y., Pui, C. H., and Downing, J. R. (2004). Gene expression profiling of pediatric acute myelogenous leukemia. *Blood* 104(12):3679–3687.

Rouillard, J. M., Herbert, C. J., and Zuker, M. (2002). OligoArray: genome-scale oligonucleotide design for microarrays. *Bioinformatics* 18(3):486–487.

Saidi, S. A., Holland, C. M., Kreil, D. P., MacKay, D. J., Charnock-Jones, D. S., Print, C. G., and Smith, S. K. (2004). Independent component analysis of microarray data in the study of endometrial cancer. *Oncogene* 23(39):6677–6683.

Sakamoto, M., Kondo, A., Kawasaki, K., Goto, T., Sakamoto, H., Miyake, K., Koyamatsu, Y., Akiya, T., Iwabuchi, H., Muroya, T., Ochiai, K., Tanaka, T., Kikuchi, Y., and Tenjin, Y. (2001). Analysis of gene expression profiles associated with cisplatin resistance in human ovarian cancer cell lines and tissues using cDNA microarray. *Hum. Cell* 14(4):305–315.

Sallinen, S. L., Sallinen, P. K., Haapasalo, H. K., Helin, H. J., Helen, P. T., Schraml, P., Kallioniemi, O. P., and Kononen, J. (2000). Identification of differentially expressed genes in human gliomas by DNA microarray and tissue chip techniques. *Cancer Res.* 60(23):6617–6622.

Sanchez-Carbayo, M. (2004). Recent advances in bladder cancer diagnostics. *Clin. Biochem.* 37(7):562–571.

Sanchez-Carbayo, M., Socci, N. D., Charytonowicz, E., Lu, M., Prystowsky, M., Childs, G., and Cordon-Cardo, C. (2002). Molecular profiling of bladder cancer using cDNA microarrays: defining histogenesis and biological phenotypes. *Cancer Res.* 62(23):6973–6980.

Sanchez-Carbayo, M., Capodieci, P., and Cordon-Cardo, C. (2003a). Tumor suppressor role of KiSS-1 in bladder cancer: loss of KiSS-1 expression is associated with bladder cancer progression and clinical outcome. *Am. J. Pathol.* 162(2):609–617.

Sanchez-Carbayo, M., Socci, N. D., Lozano, J. J., Li, W., Charytonowicz, E., Belbin, T. J., Prystowsky, M. B., Ortiz, A. R., Childs, G., and Cordon-Cardo, C. (2003b). Gene discovery in bladder cancer progression using cDNA microarrays. *Am. J. Pathol.* 163(2):505–516.

Santin, A. D., Zhan, F., Bellone, S., Palmieri, M., Cane, S., Bignotti, E., Anfossi, S., Gokden, M., Dunn, D., Roman, J. J., O'Brien, T. J., Tian, E., Cannon, M. J., Shaughnessy, J., and Pecorelli, S. (2004a). Gene expression profiles in primary ovarian serous papillary tumors and normal ovarian epithelium: identification of candidate molecular markers for ovarian cancer diagnosis and therapy. *Int. J. Cancer* 112(1):14–25.

Santin, A. D., Zhan, F., Bellone, S., Palmieri, M., Cane, S., Gokden, M., Roman, J. J., O'Brien, T. J., Tian, E., Cannon, M. J., Shaughnessy, J., and Pecorelli, S. (2004b). Discrimination between uterine serous papillary carcinomas and ovarian serous papillary tumours by gene expression profiling. *Br. J. Cancer* 90(9):1814–1824.

Sasik, R., Hwa, T., Iranfar, N., and Loomis, W. F. (2001). Percolation clustering: a novel algorithm applied to the clustering of gene expression patterns in dictyostelium development. *Pacific Symposium on Biocomputing* 6:335–347. (Available online at http://psb.stanford.edu.)

Savage, K. J., Monti, S., Kutok, J. L., Cattoretti, G., Neuberg, D., De Leval L., Kurtin, P., Dal Cin, P., Ladd, C., Feuerhake, F., Aguiar, R. C., Li, S., Salles, G., Berger, F., Jing, W., Pinkus, G. S., Habermann, T., Dalla-Favera, R., Harris, N. L., Aster, J. C., Golub, T. R., and Shipp, M. A. (2003). The molecular signature of mediastinal large B-cell lymphoma differs from that of other diffuse large B-cell lymphomas and shares features with classical Hodgkin lymphoma. *Blood* 102(12):3871–3879.

Sawiris, G. P., Sherman-Baust, C. A., Becker, K. G., Cheadle, C., Teichberg, D., and Morin, P. J. (2002). Development of a highly specialized cDNA array for the study and diagnosis of epithelial ovarian cancer. *Cancer Res.* 62(10):2923–2928.

Schadt, E. E., Li, C., Su, C., and Wong, W. H. (2000). Analyzing high-density oligonucleotide gene expression array data. *J. Cell. BioChem.* 80:192–201.

Schaner, M. E., Ross, D. T., Ciaravino, G., Sorlie, T., Troyanskaya, O., Diehn, M., Wang, Y. C., Duran, G. E., Sikic, T. L., Caldeira, S., Skomedal, H., Tu, I. P., Hernandez-Boussard, T., Johnson, S. W., O'Dwyer, P. J., Fero, M. J., Kristensen, G. B., Borresen-Dale, A. L., Hastie, T., Tibshirani, R., van de Rijn, M., Teng, N. N., Longacre, T. A., Botstein, D., Brown, P. O., and Sikic, B. I. (2003). Gene expression patterns in ovarian carcinomas. *Mol. Biol. Cell* 14(11):4376–4386.

Schena, M. (1999). *DNA Microarrays: A Practical Approach* (Practical Approach Series, 205). Oxford, UK: Oxford University Press.

Schena, M. (2000). *Microarray Biochip Technology*. Sunnyvale, CA: Eaton.

Scherf, M., Klingenhoff, A., and Werner, T. (2000). Highly specific localization of promoter regions in large genomic sequences by PromoterInspector: a novel context analysis approach. *J. Mol. Biol.* 297:599–606.

Schoch, C., Kohlmann, A., Schnittger, S., Brors, B., Dugas, M., Mergenthaler, S., Kern, W., Hiddemann, W., Eils, R., and Haferlach, T. (2002). Acute myeloid leukemias with reciprocal rearrangements can be distinguished by specific gene expression profiles. *Proc. Natl. Acad. Sci. USA* 99(15):10008–10013. Epub 2002 Jul 08.

Schuchhardt, J., Beule, D., Malik, A., Wolski, E., Eickhoff, H., Lehrach, H., and Herzel, H. (2000). Normalization strategies for cDNA microarrays. *Nucleic Acids Res.* 28:e47.

Schwartz, D. R., Kardia, S. L., Shedden, K. A., Kuick, R., Michailidis, G., Taylor, J. M., Misek, D. E., Wu, R., Zhai, Y., Darrah, D. M., Reed, H., Ellenson, L. H., Giordano, T. J., Fearon, E. R., Hanash, S. M., and Cho, K. R. (2002). Gene expression in ovarian cancer reflects both morphology and biological behavior, distinguishing clear cell from other poor-prognosis ovarian carcinomas. *Cancer Res.* 62(16):4722–4729.

Seftor, E. A., Meltzer, P. S., Kirschmann, D. A., Pe'er, J., Maniotis, A. J., Trent, J. M., Folberg, R., and Hendrix, M. J. (2002). Molecular determinants of human uveal melanoma invasion and metastasis. *Clin. Exp. Metastasis* 19(3):233–246.

Segal, N. H., Pavlidis, P., Noble, W. S., Antonescu, C. R., Viale, A., Wesley, U. V., Busam, K., Gallardo, H., DeSantis, D., Brennan, M. F., Cordon-Cardo, C., Wolchok, J. D., and Houghton, A. N. (2003). Classification of clear-cell sarcoma as a subtype of melanoma by genomic profiling. *J. Clin. Oncol.* 21(9):1775–1781.

Shai, R., Shi, T., Kremen, T. J., Horvath, S., Liau, L. M., Cloughesy, T. F., Mischel, P. S., and Nelson, S. F. (2003). Gene expression profiling identifies molecular subtypes of gliomas. *Oncogene* 22(31):4918–4923.

Sheng, Q., Moreau, Y., and De Moor, B. (2003). Biclustering microarray data by Gibbs sampling. *Bioinformatics* 19(Suppl 2):II196–II205.

Shipp, M. A., Ross, K. N., Tamayo, P., Weng, A. P., Kutok, J. L., Aguiar, R. C., Gaasenbeek, M., Angelo, M., Reich, M., Pinkus, G. S., Ray, T. S., Koval, M. A., Last, K. W., Norton, A., Lister, T. A., Mesirov, J., Neuberg, D. S., Lander, E. S., Aster, J. C., and Golub, T. R. (2002). Diffuse large B-cell lymphoma outcome prediction by gene-expression profiling and supervised machine learning. *Nature Med.* 8(1):68–74.

Shmulevich, I., Hunt, K., El-Naggar, A., Taylor, E., Ramdas, L., Laborde, P., Hess, K. R., Pollock, R., and Zhang, W. (2002). Tumor specific gene expression profiles in human leiomyosarcoma: an evaluation of intratumor heterogeneity. *Cancer* 94(7):2069–2075.

Simon, R., Radmacher, M. D., Dobbin, K., and McShane, L. M. (2003). Pitfalls in the use of DNA microarray data for diagnostic and prognostic classification. *J. Natl. Cancer Inst.* 95(1):14–18 (review).

Singh, D., Febbo, P. G., Ross, K., Jackson, D. G., Manola, J., Ladd, C., Tamayo, P., Renshaw, A. A., D'Amico, A. V., Richie, J. P., Lander, E. S., Loda, M., Kantoff, P. W., Golub, T. R., and Sellers, W. R. (2002). Gene expression correlates of clinical prostate cancer behavior. *Cancer Cell* 1(2):203–209.

Singh-Gasson, S., Green, R. D., Yue, Y., Nelson, C., Blattner, F., Sussman, M. R., and Cerrina F. (1999). Maskless fabrication of light-directed oligonucleotide microarrays using a digital micromirror array. *Nature Biotechnol.* 17:974–978.

Skovgaard, M., Jensen, L. J., Brunak, S., Ussery, D., and Krogh, A. (2001). On the total number of genes and their length distribution in complete microbial genomes. *Trends Genet.* 17:425–428.

Sorlie, T., Perou, C. M., Tibshirani, R., Aas, T., Geisler, S., Johnsen, H., Hastie, T., Eisen, M. B., van de Rijn, M., Jeffrey, S. S., Thorsen, T., Quist, H., Matese, J. C., Brown, P. O., Botstein, D., Eystein, L. O., and Borresen-Dale, A. L. (2001). Gene expression patterns of breast carcinomas distinguish tumor subclasses with clinical implications. *Proc. Natl. Acad. Sci. USA* 98(19):10869–10874.

Sorlie, T., Tibshirani, R., Parker, J., Hastie, T., Marron, J. S., Nobel, A., Deng, S., Johnsen, H., Pesich, R., Geisler, S., Demeter, J., Perou, C. M., Lonning, P. E., Brown, P. O., Borresen-Dale, A. L., and Botstein, D. (2003). Repeated observation of breast tumor subtypes in independent gene expression data sets. *Proc. Natl. Acad. Sci. USA* 100(14):8418–8423.

Sotiriou, C., Neo, S. Y., McShane, L. M., Korn, E. L., Long, P. M., Jazaeri, A., Martiat, P., Fox, S. B., Harris, A. L., and Liu, E. T. (2003). Breast cancer classification and prognosis based on gene expression profiles from a population-based study. *Proc. Natl. Acad. Sci. USA* 100(18):10393–10398. Epub 2003 Aug 13.

Spellman, P., Sherlock, G., Zhang, M., Lyer, V., Anders, K., Eisen, M., Brown, P., Botstein, D., and Futcher, B. (1998). Comprehensive identification of cell cycle-regulated genes of yeast *S. cerevisiae* by microarray hybridization. *Mol. Biol. Cell* 9:3273–3297.

Spentzos, D., Levine, D. A., Kolia, S., Otu, H., Boyd, J., Libermann, T. A., and Cannistra, S. A. (2005). Unique gene expression profile based on pathologic response in epithelial ovarian cancer. *J. Clin. Oncol.* 23(31):7911–7918.

Spicker, J. S., Wikman, F, Lu M. L., Cordon-Cardo C., Workman, C., Ørntoft, T. F., Brunak, S., and Knudsen, S. (2002). Neural network predicts sequence of TP53 gene based on DNA chip. *Bioinformatics* 18:1133–1134.

Staal, F. J., van der Burg, M., Wessels, L. F., Barendregt, B. H., Baert, M. R., van den Burg, C. M., van Huffel, C., Langerak, A. W., van der Velden, V. H., Reinders, M. J., and van Dongen, J. J. (2003). DNA microarrays for comparison of gene expression profiles between diagnosis and relapse in precursor-B acute lymphoblastic leukemia: choice of technique and purification influence the identification of potential diagnostic markers. *Leukemia* 17(7):1324–1332. Erratum in: *Leukemia* 2004 May;18(5):1041.

Stegmaier, K., Corsello, S. M., Ross, K. N., Wong, J. S., Deangelo, D. J., and Golub, T. R. (2005). Gefitinib (Iressa) induces myeloid differentiation of acute myeloid leukemia. *Blood* 106(8):2841–2848.

Stratowa, C., Loffler, G., Lichter, P., Stilgenbauer, S., Haberl, P., Schweifer, N., Dohner, H., and Wilgenbus, K. K. (2001). CDNA microarray gene expression analysis of B-cell chronic lymphocytic leukemia proposes potential new prognostic markers involved in lymphocyte trafficking. *Int. J. Cancer* 91(4):474–480.

Stuart, R. O., Wachsman, W., Berry, C. C., Wang-Rodriguez, J., Wasserman, L., Klacansky, I., Masys, D., Arden, K., Goodison, S., McClelland, M., Wang, Y., Sawyers, A., Kalcheva, I., Tarin, D., and Mercola, D. In silico dissection of cell-type-associated patterns of gene expression in prostate cancer. *Proc. Natl. Acad. Sci. USA* 101(2):615–620.

Sun, Z., Yang, P., Aubry, M. C., Kosari, F., Endo, C., Molina, J., and Vasmatzis, G. (2004). Can gene expression profiling predict survival for patients with squamous cell carcinoma of the lung? *Mol. Cancer* 3(1):35–45.

Sutton, A. J., Abrams, K. R., Jones, D. J., Sheldon, T. A., and Song, F. J. (2000). *Methods for Meta-Analysis in Medical Research.* Hoboken, NJ: Wiley.

Talbot, S. G., Estilo, C., Maghami, E., Sarkaria, I. S., Pham, D. K., O-charoenrat, P., Socci, N. D., Ngai, I., Carlson, D., Ghossein, R., Viale, A., Park, B. J., Rusch, V. W., and Singh, B. (2005). Gene expression profiling allows distinction between primary and metastatic squamous cell carcinomas in the lung. *Cancer Res.* 65(8):3063–3071.

Tamayo, P., Slonim, D., Mesirov, J., Zhu, Q., Kitareewan, S., Dmitrovsky, E., Lander, E. S., and Golub, T. R. (1999). Interpreting patterns of gene expression with self-organizing maps: methods and application to hematopoietic differentiation. *Proc. Natl. Acad. Sci. USA* 96:2907–2912.

Tan, Y., Shi, L., Tong, W., Gene, H., and Wang, C. (2004). Multi-class tumor classification by discriminant partial least squares using microarray gene expression data and assessment of classification models. *Comput. Biol. Chem.* 28(3):235–244.

Tapper, J., Kettunen, E., El-Rifai, W., Seppala, M., Andersson, L. C., and Knuutila, S. (2001). Changes in gene expression during progression of ovarian carcinoma. *Cancer Genet. Cytogenet.* 128(1):1–6.

Teuffel, O., Dettling, M., Cario, G., Stanulla, M., Schrappe, M., Buhlmann, P., Niggli, F. K., and Schafer, B. W. (2004). Gene expression profiles and risk stratification in childhood acute lymphoblastic leukemia. *Haematologica* 89(7):801–808.

Theilhaber, J., Bushnell, S., Jackson, A., and Fuchs, R. (2001). Bayesian estimation of fold-changes in the analysis of gene expression: the PFOLD algorithm. *J. Computat. Biol.* 8(6):585–614.

Thomas, J. G., Olson, J. M., Tapscott, S. J., and Zhao, L. P. (2001). An efficient and robust statistical modeling approach to discover differentially expressed genes using genomic expression profiles. *Genome Res.* 11:1227–1236.

Thykjaer, T., Workman, C., Kruhøffer, M., Demtröder, K., Wolf, H., Andersen, L. D., Frederiksen, C. M., Knudsen, S., and Ørntoft, T. F. (2001). Identification of gene expression patterns in superficial and invasive human bladder cancer. *Cancer Res.* 61(6):2492–2499.

Tibshirani, R. (2005). Immune signatures in follicular lymphoma. *N. Engl. J. Med.* 352(14):1496–1497.

Tibshirani, R., Walther, G., Botstein, D., and Brown, P. (2000). Cluster validation by prediction strength. Technical report. Statistics Department, Stanford University. (Manuscript available at http://www-stat.stanford.edu/~tibs/research.html.)

Tonin, P. N., Hudson, T. J., Rodier, F., Bossolasco, M., Lee, P. D., Novak, J., Manderson, E. N., Provencher, D., and Mes-Masson, A. M. (2001). Microarray analysis of gene expression mirrors the biology of an ovarian cancer model. *Oncogene* 20(45):6617–6626.

Tschentscher, F., Husing, J., Holter, T., Kruse, E., Dresen, I. G., Jockel, K. H., Anastassiou, G., Schilling, H., Bornfeld, N., Horsthemke, B., Lohmann, D. R., and Zeschnigk, M. (2003). Tumor classification based on gene expression profiling shows that uveal melanomas with and without monosomy 3 represent two distinct entities. *Cancer Res.* 63(10):2578–2584.

Tsui, N. B., Ng, E. K., and Lo, Y. M. (2002). Stability of endogenous and added RNA in blood specimens, serum, and plasma. *Clin. Chem.* 48(10):1647–1653.

Tusher, V. G., Tibshirani, R., and Chu, G. (2001). Significance analysis of microarrays applied to the ionizing radiation response. *Proc. Natl. Acad. Sci. USA* 98:5119–5121. (Software available for download at http://www-stat.stanford.edu/~tibs/SAM/index.html.)

Valet, G. K., and Hoeffkes, H. G. (2004). Data pattern analysis for the individualised pretherapeutic identification of high-risk diffuse large B-cell lymphoma (DLBCL) patients by cytomics. *Cytometry* 59A(2):232–236.

Valk, P. J., Verhaak, R. G., Beijen, M. A., Erpelinck, C. A., Barjesteh van Waalwijk van Doorn-Khosrovani, S., Boer, J. M., Beverloo, H. B., Moorhouse, M. J., van der Spek, P. J., Lowenberg, B., and Delwel, R. (2004). Prognostically useful gene-expression profiles in acute myeloid leukemia. *N. Engl. J. Med.* 350(16):1617–1628.

van de Vijver, M. J., He, Y. D., van't Veer, L. J., Dai, H., Hart, A. A., Voskuil, D. W., Schreiber, G. J., Peterse, J. L., Roberts, C., Marton, M. J., Parrish, M., Atsma, D., Witteveen, A., Glas, A., Delahaye, L., van der Velde, T., Bartelink, H., Rodenhuis, S., Rutgers, E. T., Friend, S. H., and Bernards, R. (2002). A gene-expression signature as a predictor of survival in breast cancer. *N. Engl. J. Med.* 347(25):1999–2009.

van Delft, F. W., Bellotti, T., Luo, Z., Jones, L. K., Patel, N., Yiannikouris, O., Hill, A. S., Hubank, M., Kempski, H., Fletcher, D., Chaplin, T., Foot, N., Young, B. D., Hann, I. M., Gammerman, A., and Saha, V. (2005). Prospective gene expression analysis accurately subtypes acute leukaemia in children and establishes a commonality between hyperdiploidy and t(12;21) in acute lymphoblastic leukaemia. *Br. J. Haematol.* 130(1):26–35.

van den Boom, J., Wolter, M., Kuick, R., Misek, D. E., Youkilis, A. S., Wechsler, D. S., Sommer, C., Reifenberger, G., and Hanash, S. M. (2003). Characterization of gene expression profiles associated with glioma progression using oligonucleotide-based microarray analysis and real-time reverse transcription-polymerase chain reaction. *Am. J. Pathol.* 163(3):1033–1043.

van der Velden, P. A., Zuidervaart, W., Hurks, M. H., Pavey, S., Ksander, B. R., Krijgsman, E., Frants, R. R., Tensen, C. P., Willemze, R., Jager, M. J., and Gruis, N. A. (2003). Expression profiling reveals that methylation of TIMP3 is involved in uveal melanoma development. *Int. J. Cancer* 106(4):472–479.

van Kampen, A. H., van Schaik, B. D., Pauws, E., Michiels, E. M., Ruijter, J. M., Caron, H. N., Versteeg, R., Heisterkamp, S. H., Leunissen, J. A., Baas, F., and van der Mee M. (2000). USAGE: a web-based approach towards the analysis of SAGE data. *Bioinformatics* 16:899–905.

van't Veer, L. J., Dai, H., van de Vijver, M. J., He, Y. D., Hart, A. A., Mao, M., Peterse, H. L., van der Kooy, K., Marton, M. J., Witteveen, A. T., Schreiber, G. J., Kerkhoven, R. M., Roberts, C., Linsley, P. S., Bernards, R., and Friend, S. H. (2002). Gene expression profiling predicts clinical outcome of breast cancer. *Nature* 415:530–536.

van't Veer, L. J., Dai, H., van de Vijver, M. J., He, Y. D., Hart, A. A., Bernards, R., and Friend, S. H. (2003). Expression profiling predicts outcome in breast cancer. *Breast Cancer Res.* 5(1):57–58.

Varotto, C., Richly, E., Salamini, F., and Leister, D. (2001). GST-PRIME: a genome-wide primer design software for the generation of gene sequence tags. *Nucleic Acids Res.* 29:4373–4377.

Velculescu, V. E., Zhang, L., Vogelstein, B., and Kinzler, K. W. (1995). Serial analysis of gene expression. *Science* 270:484–487.

Venables, W. N., and Ripley, B. D. (1999). *Modern Applied Statistics with S-PLUS*, 3rd ed. New York: Springer.

Venet, D., Pecasse, F., Maenhaut, C., and Bersini, H. (2001). Separation of samples into their constituents using gene expression data. *Bioinformatics* 17(Suppl 1):S279–S287.

Vingron, M. (2001). Bioinformatics needs to adopt statistical thinking (editorial). *Bioinformatics* 17:389–390.

Virtanen, C., Ishikawa, Y., Honjoh, D., Kimura, M., Shimane, M., Miyoshi, T., Nomura, H., and Jones, M. H. (2002). Integrated classification of lung tumors and cell lines by expression profiling. *Proc. Natl. Acad. Sci. USA* 99(19):12357–12362.

von Heydebreck, A., Huber, W., Poustka, A., and Vingron, M. (2001). Identifying splits with clear separation: a new class discovery method for gene expression data. *Bioinformatics* 17(Suppl 1):S107–S114.

Wahde, M., Klus, G. T., Bittner, M. L., Chen, Y., and Szallasi, Z. (2002). Assessing the significance of consistently mis-regulated genes in cancer associated gene expression matrices. *Bioinformatics* 18(3):389-394.

Wall, M. E., Dyck, P. A., and Brettin, T. S. (2001). SVDMAN—singular value decomposition analysis of microarray data. *Bioinformatics* 17:566–568.

Wang, E., Miller, L. D., Ohnmacht, G. A., Mocellin, S., Perez-Diez, A., Petersen, D., Zhao, Y., Simon, R., Powell, J. I., Asaki, E., Alexander, H. R., Duray, P. H., Herlyn, M., Restifo, N. P., Liu, E. T., Rosenberg, S. A., and Marincola, F. M. (2002). Prospective molecular profiling of melanoma metastases suggests classifiers of immune responsiveness. *Cancer Res.* 62(13):3581–3586.

Wang, J., Coombes, K. R., Highsmith, W. E., Keating, M. J., and Abruzzo, L. V. (2004). Differences in gene expression between B-cell chronic lymphocytic leukemia and normal B cells: a meta-analysis of three microarray studies. *Bioinformatics* 20(17):3166–3178.

Wang, K., Gan, L., Jeffery, E., Gayle, M., Gown, A. M., Skelly, M., Nelson, P. S., Ng, W. V., Schummer, M., Hood, L., and Mulligan, J. (1999). Monitoring gene expression profile changes in ovarian carcinomas using cDNA microarray. *Gene* 229(1-2):101–108.

Wang, Y., Jatkoe, T., Zhang, Y., Mutch, M. G., Talantov, D., Jiang, J., McLeod, H. L., and Atkins, D. (2004). Gene expression profiles and molecular markers to predict recurrence of Dukes' B colon cancer. *J. Clin. Oncol.* 22(9):1564–1571.

Wang, Y., Klijn, J. G., Zhang, Y., Sieuwerts, A. M., Look, M. P., Yang, F., Talantov, D., Timmermans, M., Meijer-van, G. E., Yu, J., Jatkoe, T., Berns, E. M., Atkins, D., and Foekens, J. A. (2005). Gene-expression profiles to predict distant metastasis of lymph-node-negative primary breast cancer. *Lancet* 365(9460):671–679.

Wang, Z. C., Lin, M., Wei, L. J., Li, C., Miron, A., Lodeiro, G., Harris, L., Ramaswamy, S., Tanenbaum, D. M., Meyerson, M., Iglehart, J. D., and Richardson, A. (2004). Loss of heterozygosity and its correlation with expression profiles in subclasses of invasive breast cancers. *Cancer Res.* 64(1):64–71.

Watson, M. A., Perry, A., Budhjara, V., Hicks, C., Shannon, W. D., and Rich, K. M. (2001). Gene expression profiling with oligonucleotide microarrays distinguishes World Health Organization grade of oligodendrogliomas. *Cancer Res.* 61(5):1825–1829.

Wei, J. S., Greer, B. T., Westermann, F., Steinberg, S. M., Son, C. G., Chen, Q. R., Whiteford, C. C., Bilke, S., Krasnoselsky, A. L., Cenacchi, N., Catchpoole, D., Berthold, F., Schwab, M., and Khan, J. (2004). Prediction of clinical outcome using gene expression profiling and artificial neural networks for patients with neuroblastoma. *Cancer Res.* 64(19):6883–6891.

Weigelt, B., Glas, A. M., Wessels, L. F., Witteveen, A. T., Peterse, J. L., and van't Veer, L. J. (2003). Gene expression profiles of primary breast tumors maintained in distant metastases. *Proc. Natl. Acad. Sci. USA* 100(26):15901–15905.

Weigelt, B., Wessels, L. F., Bosma, A. J., Glas, A. M., Nuyten, D. S., He, Y. D., Dai, H., Peterse, J. L., and Van't Veer, L. (2005). No common denominator for breast cancer lymph node metastasis. *Br. J. Cancer* 93(8):924–932.

Welsh, J. B., Sapinoso, L. M., Su, A. I., Kern, S. G., Wang-Rodriguez, J., Moskaluk, C. A., Frierson, H. F., and Hampton, G. M. (2001a). Analysis of gene expression identifies candidate markers and pharmacological targets in prostate cancer. *Cancer Res.* 61(16):5974–5978.

Welsh, J. B., Zarrinkar, P. P., Sapinoso, L. M., Kern, S. G., Behling, C. A., Monk, B. J., Lockhart, D. J., Burger, R. A., and Hampton, G. M. (2001b). Analysis of gene expression profiles in normal and neoplastic ovarian tissue samples identifies candidate molecular markers of epithelial ovarian cancer. *Proc. Natl. Acad. Sci. USA* 98(3):1176–1181.

West, M., Blanchette, C., Dressman, H., Huang, E., Ishida, S., Spang, R., Zuzan, H., Olson, J. A., Marks, J. R., and Nevins, J. R. Predicting the clinical status of human breast cancer by using gene expression profiles. *Proc. Natl. Acad. Sci. USA* 98(20):11462–11467.

Wigle, D., Jurisica, I., Radulovich, N., Pintilie, M., Rossant, J., Liu, N., Lu, C., Woodgett, J., Seiden, I., Johnston, M., Keshavjee, S., Darling, G., Winton, T., Breitkreutz, B., Jorgenson, P., Tyers, M., Shepherd, F. A., and Tsao, M. S. (2002). Molecular profiling of non-small cell lung cancer and correlation with disease-free survival *Cancer Res.* 62(11):3005-3008.

Wikman, F. P., Lu, M. L., Thykjaer, T., Olesen, S. H., Andersen, L. D., Cordon-Cardo, C., and Ørntoft, T. F. (2000). Evaluation of the performance of a p53 sequencing microarray chip using 140 previously sequenced bladder tumor samples. *Clin. Chem.* 46:1555–1561.

Willenbrock, H., Juncker, A. S., Schmiegelow, K., Knudsen, S., and Ryder, L. P. (2004). Prediction of immunophenotype, treatment response, and relapse in childhood acute lymphoblastic leukemia using DNA microarrays. *Leukemia* 18(7):1270–1277.

Wodicka, L., Dong, H., Mittmann, M., Ho, M. H., and Lockhart, D. J. (1997). Genome-wide expression monitoring in *Saccharomyces cerevisiae. Nature Biotechnol.* 15:1359–1367.

Wolfinger, R. D., Gibson, G., Wolfinger, E. D., Bennett, L., Hamadeh, H., Bushel, P., Afshari, C., and Paules, R. S. (2001). Assessing gene significance from cDNA microarray expression data via mixed models. *J. Computat. Biol.* 8(6):625–637.

Wolfsberg, T. G., Gabrielian, A. E., Campbell, M. J., Cho, R. J., Spouge, J. L., and Landsman, D. (1999). Candidate regulatory sequence elements for cell cycle-dependent transcription in *Saccharomyces cerevisiae. Genome Res.* 9:775–792.

Wong, K. K., Cheng, R. S., and Mok, S. C. (2001). Identification of differentially expressed genes from ovarian cancer cells by MICROMAX cDNA microarray system. *Biotechniques* 30(3):670–675.

Workman, C., and Stormo, G. D. (2000) ANN-Spec: a method for discovering transcription factor binding sites with improved specificity. *Pacific Symposium on Biocomputing 2000*, 467–478. (Available online at http://psb.stanford.edu.)

Workman, C., Jensen, L. J., Jarmer, H., Berka, R., Saxild, H. H., Gautier, L., Nielsen, C., Nielsen, H. B., Brunak, S., and Knudsen, S. (2002) A new non-linear normalization method for reducing variance between DNA microarray experiments. *Genome Biol.* 3(9):0048. (Software available in affy package of Bioconductor http://www.bioconductor.org.)

Wright, G., Tan, B., Rosenwald, A., Hurt, E. H., Wiestner, A., and Staudt, L. M. (2003). A gene expression-based method to diagnose clinically distinct subgroups of diffuse large B cell lymphoma. *Proc. Natl. Acad. Sci. USA* 100(17):9991–9996.

Xia, X, and Xie, Z. (2001). AMADA: analysis of microarray data. *Bioinformatics* 17:569–570.

Xing, E. P., and Karp, R. M. (2001). CLIFF: clustering of high-dimensional microarray data via iterative feature filtering using normalized cuts. *Bioinformatics* 17(Suppl 1):S306–S315.

Xiong, M., Jin, L., Li, W., and Boerwinkle, E. (2000). Computational methods for gene expression-based tumor classification. *Biotechniques* 29:1264–1268.

Xu, J., Stolk, J. A., Zhang, X., Silva, S. J., Houghton, R. L., Matsumura, M., Vedvick, T. S., Leslie, K. B., Badaro, R., and Reed, S. G. (2000). Identification of differentially expressed genes in human prostate cancer using subtraction and microarray. *Cancer Res.* 60(6):1677–1682.

Xu, S., Mou, H., Lu, G., Zhu, C., Yang, Z., Gao, Y., Lou, H., Liu, X., Cheng, Y., and Yang, W. (2002). Gene expression profile differences in high and low metastatic human ovarian cancer cell lines by gene chip. *Chin. Med. J. (Engl.)* 115(1):36–41.

Yagi, T., Morimoto, A., Eguchi, M., Hibi, S., Sako, M., Ishii, E., Mizutani, S., Imashuku, S., Ohki, M., and Ichikawa, H. (2003). Identification of a gene expression signature associated with pediatric AML prognosis. *Blood* 102(5):1849–1856. Epub 2003 May 08.

Yamagata, N., Shyr, Y., Yanagisawa, K., Edgerton, M., Dang, T. P., Gonzalez, A., Nadaf, S., Larsen, P., Roberts, J. R., Nesbitt, J. C., Jensen, R., Levy, S., Moore, J. H., Minna, J. D., and Carbone, D. P. (2003). A training-testing approach to the molecular classification of resected non-small cell lung cancer. *Clin. Cancer Res.* 9(13):4695–4704.

Yamamoto, M., Wakatsuki, T., Hada, A., and Ryo, A. (2001). Use of serial analysis of gene expression (SAGE) technology. *J. Immunol. Methods* 250:45–66 (review).

Yang, P., Sun, Z., Aubry, M. C., Kosari, F., Bamlet, W., Endo, C., Molina, J. R., and Vasmatzis, G. (2004). Study design considerations in clinical outcome research of lung cancer using microarray analysis. *Lung Cancer* 46(2):215–226.

Yang, Y. H., Buckley, M. J., and Speed, T. P. (2001a). Analysis of cDNA microarray images. *Briefings in Bioinformatics* 2(4):341–349.

Yang, Y. H., Buckley, M. J., Dudoit, S., and Speed, T. P. (2001b). Comparison of methods for image analysis on cDNA microarray data. Technical report #584, Department of Statistics, University of California, Berkeley.

Yang, Y. H., Dudoit, S., Luu, P., Lin, D. M., Peng, V., Ngai, J., and Speed, T. P. (2002). Normalization for cDNA microarray data: a robust composite method addressing single and multiple slide systematic variation. *Nucleic Acids Res.* 30(4):e15.

Yasojima, K., McGeer, E. G., and McGeer, P. L. (2001). High stability of mRNAs postmortem and protocols for their assessment by RT-PCR. *Brain Res. Brain Res. Protoc.* 8(3):212–218.

Yeoh, E. J., Ross, M. E., Shurtleff, S. A., Williams, W. K., Patel, D., Mahfouz, R., Behm, F. G., Raimondi, S. C., Relling, M. V., Patel, A., Cheng, C., Campana, D., Wilkins, D., Zhou, X., Li, J., Liu, H., Pui, C. H., Evans, W. E., Naeve, C., Wong, L., and Downing, J. R. (2002). Classification, subtype discovery, and prediction of outcome in pediatric acute lymphoblastic leukemia by gene expression profiling. *Cancer Cell* 1(2):133–143.

Yeung, K. Y., Fraley, C., Murua, A., Raftery, A. E., and Ruzzo, W. L. (2001a). Model-based clustering and data transformations for gene expression data. *Bioinformatics* 17:977–987.

Yeung, K. Y., Haynor, D. R., and Ruzzo, W. L. (2001b). Validating clustering for gene expression data. *Bioinformatics* 17:309–318.

Ying-Hao, S., Qing, Y., Lin-Hui, W., Li, G., Rong, T., Kang, Y., Chuan-Liang, X., Song-Xi, Q., Yao, L., Yi, X., and Yu-Ming, M. (2002). Monitoring gene expression profile changes in bladder transitional cell carcinoma using cDNA microarray. *Urol. Oncol.* 7(5):207–212.

Zhang, L., Miles, M. F., and Aldape, K. D. (2003a). A model of molecular interactions on short oligonucleotide microarrays. *Nature Biotechnol.* 21(7):818–821.

Zhang, L., Miles, M. F., and Aldape, K. D. (2003b). Corrigendum: a model of molecular interactions on short oligonucleotide microarrays. *Nature Biotechnol.* 21(8):941.

Zhao, L. P., Prentice, R., and Breeden, L. (2001). Statistical modeling of large microarray data sets to identify stimulus–response profiles. *Proc. Natl. Acad. Sci. USA* 98:5631–5636.

Zou, T. T., Selaru, F. M., Xu, Y., Shustova, V., Yin, J., Mori, Y., Shibata, D., Sato, F., Wang, S., Olaru, A., Deacu, E., Liu, T. C., Abraham, J. M., and Meltzer, S. J. (2002). Application of cDNA microarrays to generate a molecular taxonomy capable of distinguishing between colon cancer and normal colon. *Oncogene* 21(31):4855–4862.

Zien, A., Aigner, T., Zimmer, R., and Lengauer, T. (2001). Centralization: a new method for the normalization of gene expression data. *Bioinformatics* 17(Suppl 1):S323–S331.

Zuidervaart, W., van der Velden, P. A., Hurks, M. H., van Nieuwpoort, F. A., Out-Luiting, C. J., Singh, A. D., Frants, R. R., Jager, M. J., and Gruis, N. A. (2003). Gene expression profiling identifies tumour markers potentially playing a role in uveal melanoma development. *Br. J. Cancer* 89(10):1914–1919.

Index

Cancer Diagnostics with DNA Microarrays, By Steen Knudsen
Copyright © 2006 John Wiley & Sons, Inc.